ISBN 978-1-5283-0819-9
PIBN 10912950

1 MONTH OF
FREE
READING

at

www.ForgottenBooks.com

By purchasing this book you are eligible for one month membership to ForgottenBooks.com, giving you unlimited access to our entire collection of over 1,000,000 titles via our web site and mobile apps.

To claim your free month visit:
www.forgottenbooks.com/free912950

English
Français
Deutsche
Italiano
Español
Português

www.forgottenbooks.com

Mythology Photography **Fiction**
Fishing Christianity **Art** Cooking
Essays Buddhism Freemasonry
Medicine **Biology** Music **Ancient
Egypt** Evolution Carpentry Physics
Dance Geology **Mathematics** Fitness
Shakespeare **Folklore** Yoga Marketing
Confidence Immortality Biographies
Poetry **Psychology** Witchcraft
Electronics Chemistry History **Law**
Accounting **Philosophy** Anthropology
Alchemy Drama Quantum Mechanics
Atheism Sexual Health **Ancient History**
Entrepreneurship Languages Sport
Paleontology Needlework Islam
Metaphysics Investment Archaeology
Parenting Statistics Criminology
Motivational

OUTLINES

OF

MODERN GEOGRAPHY,

ON A NEW PLAN,

Carefully adapted to Youth.

WITH NUMEROUS ENGRAVINGS OF

CITIES, MANNERS, COSTUMES, AND CURIOSITIES.

Accompanied by an Atlas.

BY REV. CHARLES A. GOODRICH.

Boston:

PUBLISHED BY CHARLES J. HENDEE,

1836.

REMARKS TO TEACHERS.

1. It is suggested that it may be found best to require a pupil who is a beginner in Geography, to recite only what is in larger type, on his first going through this book; the rest may be recited on passing through it a second time. It is also recommended that the pupil omit answering the questions on the maps printed in italics, on first going through the book.

2. The principal features, or most important parts of Geography, are in the larger type, and should be very perfectly committed to memory. The details, or more minute particulars of the subject, are in smaller type; the teacher can make such an examination of the scholar, in these, as he deems proper. It will be observed that the italic words, in the small type, embrace the leading ideas in every sentence, and indicate where and on what subject the pupil is to be interrogated.

3. The Review, at page 229, is thought to be of importance, and its use in examining pupils is particularly recommended. It may perhaps be found too extensive for young pupils, but such parts of it as the teacher thinks less important, can be omitted by them.

4. The Map of Outlines is thought to be calculated to engage the interest of pupils, and its use will establish the main points of Geography in the mind.

OUTLINES

OF

MODERN GEOGRAPHY.

INTRODUCTION.

Q. What is *Geography?*

A. Geography is a description of the earth.

Q. What is the *earth?*

A. It is a large globe, or sphere, nearly round.

Observation. The *diameter of*, or *distance through* the earth, is about 30 miles greater from east to west, than from north to south.

Q. How do you know that the earth is *round?*

A. Because it has frequently been sailed round.

This is *also proved* by the shadow of the earth, cast on the moon in an eclipse, which is always round; and from the appearance of a ship at sea, the hull, or largest part of which, disappears sooner than the mast, because of the intervening roundness of the earth.

Q. What is the distance through the *centre* of the earth?

A. About 8,000 miles.

This distance is called the diameter of the earth, which may be represented by a straight wire passing through the centre of an apple, from one side to the other.

Q. How many *miles round* is the earth?

A. About 25,000 miles.

This is called the circumference, and may be represented by a thread passing round an apple.

Q. What number of *square miles* does the earth contain?

A. About two hundred millions.

Obs. A *square mile* is a surface measuring one mile on each side. The extent, or size of a country is expressed by the number of square miles on its surface. Thus, if a country measures two miles on one side, and three miles on the other, it is as large as six squares of a mile on each side, and is said to contain six square miles.

There is much distinction between *miles square*, and *square miles.* *Ten miles square* is a surface measuring 10 miles on each side, and contains 100 squares, of a mile on each side: *Ten square miles* is a surface measuring 10 miles on one side, and one mile on the other, and contains only 10 squares of a mile on each side. The district of Columbia measures 10 miles on each side; it is therefore ten miles square, and contains 100 square miles

1*

Q. What portion of the surface is *land*?

A. Only about one quarter: the rest is water.

Astronomy.

Q. What is *astronomy*?

A. It is the science which treats of the heavenly bodies.

Astronomy explains the motions, periods, eclipses, distances, and magnitudes of the heavenly bodies.

Q. How many *kinds of heavenly bodies* may be noticed?

A. Six kinds, viz. the sun, planets, moons, asteroids, comets, and fixed stars.

SOLAR SYSTEM.

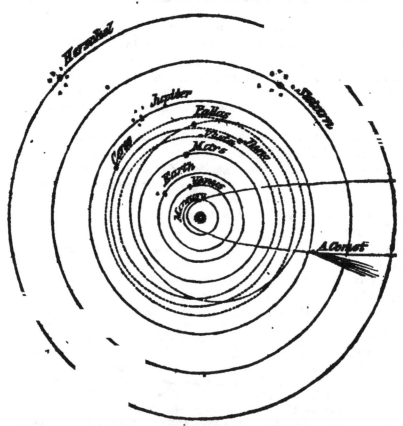

This cut represents the *Solar System;* the sun in the centre, and the planets with their moons, the asteroids, and a comet moving round him—their *paths* or *orbits* being shown by the *circular lines.*

Q. What does the *solar system* consist of?

A. Of the sun, planets, moons, asteroids, and comets.

1. This system is *called* solar, from the Latin word *sol,* which signifies the sun; and takes its name from that body, because of its importance in the system.

2. This system *supposes* that the sun is fixed in the centre, and that all the planets move round him from west to east.

SUN.

Q. What is the *sun?*

A. The sun is an immense body, which gives light and heat to the planets, moons, asteroids, and comets.

1. The sun is 883,000 miles *in diameter.*

2. It would require more than a million of *globes of the size of our earth, to make a globe of the size of the sun.*

3. It *turns upon its axis* in 25 days and 10 hours.

4. The *rotation* of the sun is known from dark spots, which appear on its surface, and which are seen for the space of twelve days and a half, and then disappear, for an equal space of time. Dr. Herschel *supposes these spots to be* the tops of mountains, seen through the luminous clouds which surround it, some of which may be 300 miles high.

5. Learned men suppose that the *light* and *heat* of the sun are either *caused,* or *excited,* by the atmosphere of the sun, but that the *body* of the sun may be solid, like our earth, and may therefore be inhabited.

PLANETS.

Q. What do you mean by the term *planet?*

A. It is derived from a Greek word, which signifies *to wander;* and is applied to certain stars, because they *move,* or *wander.*

Q. How many *kinds of planets* are there?

A. Two kinds, primary, or larger; and secondary, or smaller; to which may be added the asteroids.

Q. How many PRIMARY PLANETS are there?

A. Seven; 1 Mercury; 2 Venus; 3 Earth; 4 Mars; 5 Jupiter; 6 Saturn; 7 Herschel.

1 The planets *move round the sun*, in the order mentioned.

2. The *path* which they describe is *called* their orbit, from a Latin word, which signifies *circular*, or *round*.

3. Those planets, which are nearer the sun, move round *their orbits sooner* than those more distant, because they go faster, and have not so far to go.

4. They are *retained in their orbits*, by the joint action of two forces, called the centrip'etal and centrif'ugal forces; the *former* of which draws the planet *towards* the sun, just as powerfully as the *latter* drives it *from* the sun; so that not being able to go towards the sun, nor from it, it is compelled to move *round* it as a centre.

Q. What is the number of the SECONDARY PLANETS?

A. Eighteen.

Q. What is their *use?*

A. They accompany the larger planets, and serve to give them light.

Q. How *many* belong to each larger planet?

A. The Earth has 1; Jupiter 4; Saturn 7; Herschel 6.

Our moon is much *nearer* to the earth than any other of the heavenly bodies, being only 240,000 miles distant. With the help of a telescope, its mountains may be distinguished. The earth is about 50 times *larger* than the moon, in bulk.

ASTEROIDS.

Q. What do you mean by *asteroids?*

A. They are four very small planets, called Ceres, Pallas, Juno, and Vesta.

These planets are *situated* between the orbits of Mars and Jupiter. But little is yet known about them.

COMETS.

Q. What are *comets?*

A. They are solid bodies, which move round the sun in orbits very elliptical, or oval......*See view of solar system.*

1. The word *comet is derived from* cometa, hairy, because they generally appear with tails, resembling hair, thrown in a direction opposite to the sun.

2. These *tails are supposed by some to be* vapours, proceeding from the body of the comet by the heat of the sun. They

are of various lengths: that of 1811 was computed to be 33,000,000 of miles.

3. Their *velocity* is greater than that of the planets, and increases as they approach the sun.

4. The *number* of comets is not known. Between 400 and 500 have been observed; but only a few have been known to return, and at distant intervals.

5. Comets are probably of different *magnitudes*, but the greater number are supposed to be less than our moon.

FIXED STARS.

Q. What are the *fixed stars*?

A. They are those bodies which appear in the sky, far beyond the planets, from which they are distinguished by their *twinkling*.

1. They are called *fixed*, because they always appear in the same situation, in relation to one another.

2. Their *number* is not known. Only about 1,000 are visible at one time to the naked eye. But, by the help of a telescope, many millions can be discovered.

3. Their *distance* is immeasurable, and almost inconceivable: a cannon ball, moving at the rate of 500 miles an hour, would not reach the nearest, in 700,000 years.

4. These stars, being at such a distance, must *shine* by their own light, and hence they are *supposed to be* suns to other systems, and may each of them be accompanied by worlds, inhabited by intelligent and immortal beings.

Globes.

Q. How is the surface of the earth *best represented*?

A. By an artificial Globe.

There are two *kinds* of globes, the terrestrial, which exhibits the countries seas, &c. as they are situated on the earth; and the celestial, which exhibits the planets and fixed stars, as they appear in the heavens.

Q. What is the *axis* of the earth?

A. It is a straight line *imagined* to pass along the centre, or middle of the earth, from north to south, round which it turns once a day.

Q. What are the *two ends* of the axis called?

A. The poles; the one the *north pole*, and the other the *south pole*.

Put a straight wire through the centre of an apple, and it

will represent the axis : the *ends* of the wire will represent the poles. The following cut represents a terrestrial globe and the lines and circles upon it.

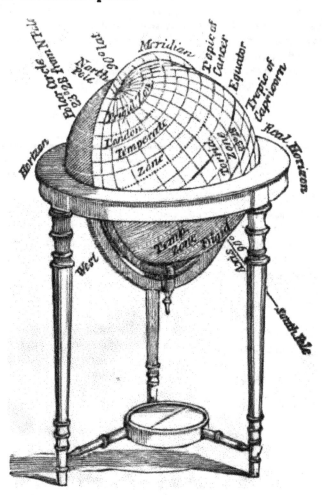

Q. What are *great* and *less circles?*

A. Great circles are those which divide the earth into two *equal* parts, or half globes; less circles divide it into two *unequal* parts.

The equator, the meridian, the ecliptic, the horizon, *are great circles;* the two tropics and the two polar circles, *are less circles.*

Q. How are circles supposed to be *divided?*

A. Into 360 equal parts, called degrees.

A *degree* on the earth is 60 geographical, or nearly 69 1-10 common miles ; *each degree is divided* into 60 minutes, and *each minute* into 60 seconds. They are *marked* thus : 24° 12′ 40″; that is, 24 degrees, 12 minutes, and 40 seconds.

Q. What is the *equator ?*

A. It is a great circle *imagined* to pass round the earth, in the middle between the poles.

Q. What are *meridians ?*

A. They are great circles, which pass through the poles, and cross the equator perpendicularly, or at right angles.

Q. What are the *tropics ?*

A. They are two less circles, drawn parallel to the equator, at the distance of 23½ degrees north and south of it.

1. The word *tropic* is *derived* from a Greek word, which signifies to turn, and *is applied to this circle*, because when the sun appears to reach it, or to be over it, he *turns* again towards the equator.

2. The tropic *north* of the equator, is *called* the tropic of Cancer ; the tropic *south*, the tropic of Capricorn.

Q. What is the *ecliptic ?*

A. It is a great circle in the heavens, in which the earth goes round the sun once a year.

Q. What are the *polar circles ?*

A. They are two large circles, parallel to the tropics, drawn at the distance of 23½ degrees from the poles.

The *northern* polar circle, is called the arctic circle ; and the *southern*, the antarctic circle.

Q. What is the *horizon ?*

A. It is the line which bounds the sight, where the earth and sky *appear* to meet.

1. This is the sensible horizon ; the *real horizon* is a great circle dividing the earth into upper and lower hemispheres. The *point directly over our heads*, is *called* the zenith, and *that under our feet*, the nadir.

2. The *four principal points* of the horizon, north, south, east, and west, are *called* Cardinal Points, from *cardo*, a hinge ; because all the intermediate points of the compass are supposed to turn on these four.

Q. What is *latitude ?*

A. It is the distance of any place, north or south

of the equator, reckoned in degrees and minutes on the meridian, and it may be 90 degrees north or south latitude.

Q. What are the *parallels* of *latitude?*

A. They are less circles, which pass round the globe, parallel to the equator.

Q. What is *longitude?*

A. It is the distance of any one place, east or west, from some particular meridian, as from that of London or Washington; and is reckoned in degrees and minutes on the *equator*, and it may be 180 degrees, east or west longitude.

Maps.

Q. What is a *Map?*

A. A map is a representation of the earth's sur face, or of some part of it, on a *level* surface.

1. In maps in general, the top is *north*, the bottom is *south*, the right hand is *east*, and the left hand *west*.

2. The *latitude* of places upon maps *is expressed by* the figures, which run up and down the sides. If the figures increase *upward*, the latitude is north; if they increase *downward*, the latitude is south.

3. The *longitude* of places upon maps *is expressed by* the figures, which run along the top and bottom. When the figures increase from *right* to *left*, the longitude is west; and when they increase from *left* to *right*, the longitude is east.

4. *Distances* upon maps are sometimes *measured* by means of a scale of miles, placed in a corner of the map.

Questions.—Map of the World.

N. B. *Before proceeding to the questions on the Maps, let the pupil, having the Map of the World before him, be taught which part of the map is north, which east—south—west. Then let him be asked as follows:—*

What line on the map represents the axis of the earth? Where is the north pole?—the south pole? Which are some of the great circles? Which are some of the small circles? What do the figures 10, 20, 30, &c. round the border of the map represent? Which is the equator? Which are meridians? Which tropics? Which is the tropic of Cancer? Which the tropic of Capricorn? How many degrees are the tropics from the equator? Which are the polar circles? What is the northern polar circle called, and what the southern? How far from the poles are they? How is latitude reckoned? How many degrees may it be? Which are lines or parallels of latitude? Why are they curved? *Ans.* To represent circles on the globe. How is longitude reckoned, and how many degrees may it be? What do the figures 10, 20, 30, &c., on the equator, represent?

Zones.

Q. What is a *Zone?*

A. It is a word literally signifying a *belt,* and denotes a portion of the earth's surface, included between certain circles.

Q. How *many* zones are there?

A. There are commonly reckoned *five;* one torrid, two temperate, and two frigid zones.

Q. What portion of the earth's surface is called the *torrid zone?*.

A. That portion which is contained between the two tropics.

1. It is *called* the *torrid,* or *burning* zone, because the sun being always over some portion of it, the weather is generally very hot.

2. The *extent* of this zone is 23½ degrees on each side of the equator, or about 3244 miles in its whole breadth, from north to south.

3. The *seasons* in this country are two, a wet season in winter, when it rains almost constantly for six months; and a dry season in summer, when rain is scarcely known.

4. The *vegetable productions* of the torrid zone are among the richest in the world, consisting of gums, spices, tea, and coffee.

5. The most remarkable *animals* are the elephant, lion, tiger, camel, besides birds of great beauty, and serpents of peculiar venom.

6. The *inhabitants* of these regions are black, or dark coloured, and are seldom distinguished for vigour, either of body or mind.

Q. What portions of the earth's surface are called the *temperate zones?*

A. Those portions which are contained between the tropics and polar circles.

1. These regions are called *temperate,* on account of the mildness of the climate, which is neither very warm, nor very cold.

2. The *extent* of each of these zones is 43 degrees in breadth, or about 2970 miles from north to south.

2

3. Each of the temperate zones has *four seasons*, Spring, Summer, Autumn, and Winter.

4. The *vegetable productions* of the temperate zones consist of the finer and coarser grains ; also cotton, vines, olives, apples, and pears.

5. *Animals* and *reptiles* dangerous to man are seldom found in these regions, but the useful and hardy animals abound.

6. The *inhabitants* of these zones are generally white, and are distinguished for genius, enterprise, and learning.

Q. What portions of the earth's surface are called the *frigid zones ?*

A. Those portions which are contained between the polar circles and the poles.

1. They are called *frigid*, because of the intense cold which prevails there.

2. The *extent* of these zones is 23½ degrees in breadth, or 1624 miles, from north to south.

3. The frigid zones have only *two seasons*, a long and intensely cold winter, and a summer, which is short and hot.

4. The *vegetable productions* are confined to a few kinds of moss.   There are no forest trees, much less any fruit trees.

5. The *animals* are of the most hardy character, such as the reindeer, and a few others, which alone are able to endure the severity of the climate.

6. The *inhabitants*, who are very few, are small in stature, with dark, or swarthy complexions, and generally possess but little intelligence.

Questions.—Map of the World.

How many zones are there ? Where is the torrid zone ? How many degrees does it extend on each side of the equator ? Which are some of the principal countries in the torrid zone ? How many temperate zones are there ? Where are they ? How many degrees does each include ? *N. B. Let the pupil count the degrees on the sides of the map.* What countries does the northern temperate zone include ? What the southern ? How many frigid zones are there ? Where are they ? How many degrees do they extend from the poles ? What land do you find in the northern frigid zone ? What in the southern ?

Natural Divisions.

Q. Which are the great *natural divisions* of the earth ?

A. Land and water.

1. LAND.

Q. What does the land *consist of ?*

A. Continents, islands, peninsulas, isthmuses, capes, and mountains.

Q. What is a *continent?*

A. It is a large portion of land, surrounded, but not divided by water.

There are *commonly reckoned* but two *continents*, the eastern and the western. New Hol'land has been called by some geographers a continent; but in this work, it will be classed among the islands.

Q. What is an *island?*

A. It is a smaller portion of land, surrounded by water; as *Great Britain, Ireland, Long Island.*

An *archipelago* (ark-e-pel'a-go) is an assemblage of islands.

Q. What is a *peninsula?*

A. A tract of land almost surrounded by water; as *Spain, Florida, the More'a in Greece.*

Q. What is an *isthmus?*

A. It is the narrow neck of land which connects a peninsula to the main land, or two parts of a continent, as the *Isthmus of Da'rien.*

Q. What is a *cape?*

A. It is a point of land extending into the sea; as *Cape Cod, Cape Horn.*

When a cape is considerably elevated above the level of the sea, it is *called* a promontory.

Q. What is a *mountain?*

A. It is a portion of land rising to a great height above the surrounding country.

When the *land rises only to a small height,* it is *called* a hill; when a *mountain emits smoke and flame,* it is *called* a volcano.

2. WATER.

Q. How is water *divided?*

A. Into oceans, seas, lakes, sounds, bays or gulfs, harbours, straits, and rivers.

Q. What is an *ocean?*

A. It is an immense collection of water, not entirely separated by land.

1. There are *commonly reckoned* five *oceans;* the Northern, Southern, Pacific, Indian, and Atlantic.

2. The *Northern* ocean *lies* around the north pole, and is connected with the Pacific by Behring's strait, about 48 miles in width, and with the Atlantic by the sea, or passage, which

separates Norway from Greenland. This ocean is supposed to extend about 3000 miles.

3. The *Southern* ocean *lies* around the south pole, extending to Cape Horn and the Cape of Good Hope.

4. The *Pacific* ocean *lies* on the west of America. It extends from Beh'ring's strait, about 8000 miles, to the limits of the Southern ocean; and from America to Asia, about 11,000 miles.

5. The *Indian* ocean *lies* between Africa and New Holland. Its extent from east to west is from 3000 to 6000 miles, and from north to south about 4000.

6. The *Atlantic* ocean *lies* between America and Europe, and is from 3000 to 4000 miles wide; and 9000 in *length* from the Northern to the Southern ocean.

Q. What is a *sea?*

A. It is a collection of water, less than an ocean, but communicates with it; as the *Mediterra'nean sea, the Baltic.*

Q. What is a *lake?*

A. It is a large collection of water in the interior of a country; as *Lake Supe'rior, Lake E'rie.*

Q. What is a *sound?*

A. It is a strait, so shallow, that it can be sounded, or measured with a lead and line; as *Long Island Sound.*

Q. What are *bays or gulfs?*

A. They are parts of an ocean, running up into the land; as *Hudson's bay, the Gulf of Mex'ico.*

Q. What is a *harbour?*

A. It is a small bay, where ships may anchor.

A *road* is a place of anchorage on an open coast.

Q. What is a *strait?*

A. It is a narrow channel connecting two large bodies of water; as the *Strait of Gibral'tar.*

Q. What is a *river?*

A. It is a large stream of water proceeding from springs, or lakes, and emptying itself into a sea or ocean.

Creeks and *rivulets* are small rivers. A *cataract* is a river, or stream of water, falling over a precipice.

Questions.—Map of the World.

How is the earth divided? Which is land? Which water? Point

out a continent—an island—an Archipelago—a peninsula—isthmus—cape—mountain.

How many oceans are there? Which are they? Where are they? Which is the largest? Which the 2d, 3d, 4th, 5th? Point out a sea—a lake—sound—bay or gulf—strait—river.

Political Divisions.

Q. Which are some of the *political divisions* of the earth?

A. Empires, kingdoms, duchies, grand duchies, principalities, provinces, and countries.

Q. What is an *empire?*

A. An empire consists of several countries, under the dominion of an *emperor*.

Q. What is a *kingdom?*

A. It consists of a single country, subject to a *king*.

Q. What is a *duchy, grand duchy*, and *principality?*

A. They are smaller portions of country, subject severally, to a *duke*, a *grand duke*, and a *prince*, who are themselves subject to some higher power.

. The smaller subdivisions of countries are *states, provinces, departments, cities*, and *towns*.

Government.

Q. What is *government?*

A. The power which makes laws, and carries them into effect.

Obs. Every nation has found it necessary to have laws, the proper objects of which are, to secure the rights of individuals, to enforce justice, and punish crimes. The manner of making and executing these laws, as well as the laws themselves, are very different, in different countries.

Q. How many *forms of government* prevail?

A. Two—monarchical, and republican.

Other forms of government have existed, such as an *aristocracy*, which is a government in the hands of a few men, usually called nobles; a *democracy*, in which the people assemble to make the laws. These kinds of government may be still found mixed with some of the existing governments, but not in a distinct form.

Q. What is a *monarchy?*

A. A monarchy is a government, in which the supreme power is vested in one man.

2*

1. An *hereditary* monarchy, is that, in which the throne descends, by inheritance, to a relation of the same family; nearly, if not all, the existing monarchies are of this kind.

2. A *limited* monarchy, is that, in which the power of the monarch is limited by a constitution, or an assembly of the people; as that of *Great Britain, France,* and others.

3. An *absolute* or *despotic* monarchy, is that, in which the power of the monarch is uncontrolled; as that of *Portugal, Persia.*

Q. What is a *republic?*

A. It is a government administered by rulers, chosen by the body of the people, for a limited time; as that of *Massachusetts, New York.*

A *confederation,* or federal republic, is a union of several independent states, for mutual aid and defence, under the direction of a congress, or general assembly; as that of the *United States.*

Religion.

Q. Into how many *classes* may the inhabitants of the earth be divided in respect to *religion?*

A. Into four classes—viz. Pagans, Mahometans, Jews, and Christians.

Q. Who are *Pagans?*

A. Those who believe in many gods, and worship idols.

1. The *objects* which Pagans worship are very numerous. Among these objects, are the sun, moon, stars—fire and water—rivers, mountains, animals—and even a shapeless block of wood and stone.

2. The *rites* which they practise to please their idols, and to obtain the forgiveness of their sins, are often very absurd. Some stand for years in the same position, until they are stiffened;—others broil before a fire, or lie on beds of spikes;—others are sawn asunder, or throw themselves under the wheels of an idol's car, and are crushed to death. By these, and similar practices, they expect not only to appease their deities, but even to become superior to them.

Q. Who are *Mahometans?*

A. Those who believe in the *Koran,* a book of Ma'homet, an Arabian impostor, who lived about 600 years after Christ, and pretended to be inspired.

1. *Mahomet* was *born* at Mecca, in Arabia. His parentage was mean, and his education neglected. But he possessed talents and cunning. In the Koran he *teaches* the worship of one God, and forbids idolatry. He *acknowledges* Christ to be a great prophet, but claims to be superior to him. He *enjoined his followers* to pray five times every day—to abstain from pork and spirituous liquors—and during the month *Ramadan*, to neither eat, drink, nor smoke, between sun-rise and sun-set; and once, at least, in their life time, to perform a pilgrimage to Mecca.

2. Mahomet *propagated* his religion by the sword, and taught that it was a crime worthy of death to profess any other.

3. He *promised* to his faithful followers a sensual paradise, hereafter.

4. *Mahometans* are sometimes *called* Mussulmans; their *priests are called* moolas, or imans; and the *chief priests* of Turkey, a mufti.

Q. Who are the *Jews?*

A. They are the descendants of Abraham, and believe in the Scriptures of the Old Testament, but not in the New Testament. They consider Christ as an impostor, but expect a Messiah yet to come.

1. The Jews *formerly dwelt* in Judea, and were a people eminently blessed by God: but for rejecting Christ they are, though still a *distinct* people, scattered abroad through all parts of the earth.

2. They are not *admitted to the common privileges of citizens in any* country in the world, except in the United States.

Q. Who are *Christians*?

A. They are those who believe in Christ as the Saviour, and receive the Scriptures of the Old and New Testaments, as the rule of their faith and practice.

Q. Which are the three principal *divisions of the Christians?*

A. Roman Catholics, Greek Christians, and Protestants.

Q. Who are the *Roman Catholics?*

A. Those who acknowledge the *Pope*, residing at Rome, as the head of the Church, and believe him to be infallible.

WESTERN CONTINENT

Q. How is the Western continent *bounded?*

A. N. by the F... o....; E. by the A....; S. it terminates in a point called c... H..; W. by the P... o...

1. The *western* continent is about 9000 miles *long,* and on an average, about 1500 miles *broad.* It is supposed to contain from 14 to 16,000,000 *square miles.*

2. It is *distinguished* for its large rivers, numerous lakes, and lofty and extended ranges of mountains.

Q. How long has this continent *been known to Europeans?*

A. About 300 years.

1. It was *discovered* in 1492, by Christopher Columbus, a native of Gen'oa, who, at that time, was on a voyage of discovery, in the service of Spain.

2. At the time of its discovery, it was *inhabited* by numerous tribes of Indians, whose descendants still possess a greater part of the soil.

Q. What is the *population* of all America supposed to be?

A. About 35,000,000.

These may be *divided* into three classes, *viz :* Whites, Negroes, and Indians. The *whites are the descendants* of Europeans; the *negroes* the descendants of Africans; and the *Indians* the descendants of those who possessed the country, when Columbus discovered it.

- Q. What remarkable range of *mountains* does this continent contain?

A. A range which runs the whole length of the continent, and which, with its windings, is supposed to be 11,500 miles long; in South America this range is called the *An'des;* in Mex'ico, the *Cordil'-leras of Mexico;* and north of Mexico, the *Rocky mountains.*

Q. What can you say of the *climate?*

A. The continent has every variety of climate; but a greater degree of cold prevails than in the same parallels in the eastern continent.

Q. How is the continent *divided?*

A. Into North and South America; between which, on the *eastern* side, the West In'dia Islands are found

The *line* which divides North from South America *crosses the isthmus of Da'rien* about seven or eight degrees, north latitude.

Questions.—Map of the World.

What ocean is situated north of America? What east? south? west? Which extends south farthest, the western or eastern continent? Through how many zones does it run? Where does the equator cross it? On which side of the equator is the most land? What islands lie far to the west of America? Into what is the western continent divided? By what? *In what latitude is the Isthmus of Darien? What sea and gulf are situated at nearly equal distances between the two extremities of the western continent?* What islands lie between North and South America? On which side? What great chain of mountains runs through the continent? On which side?

NORTH AMERICA.

North American Indians.

Q. How is North America *bounded?*

A. N. by the F... O...; E. by the A...; on the S. it is connected with South America by the I... of D...; W. by the P... O...

1. The *length* of North America is about 4500 miles, and average *breadth* 2500. It contains 7 or 8,000,000 of *square miles*, and 20,000,000 of *inhabitants.*

2. The *western* coast of North America has a mild *climate*, but the *eastern* parts are much colder than in the same latitudes in Europe.

Q. Which are the principal *mountains?*

A. The Alleghany and Rocky mountains.

1. The *Alleghany, highest peak, 6th class, is* the *eastern* range, and lies wholly within the United States. The *length* of the range is about 900 miles.

2. The *Rocky Mountains, highest peaks, 4th class,* are on the *west,* and are a continuation of the great range, which comes from South America.

Q. Which are the principal *bays* or *gulfs?*

A. Baffin's Bay, Hudson's Bay, the Gulf of St. Law'rence, the Gulf of Mex'ico, and the Gulf of Califor'nia.

Q. Which are the principal *islands?*

A. Newfoundland', Cape Breton, (Bre-toon') St. John's, Rhode Island, Long Island, and the Bermu'das.

Q. Which are the principal *lakes?*

A. Slave Lake, Win'nipeg, Superior, Hu'ron, Mich'igan, (Mish'-e-gan) E'rie, and Onta'rio.

Lake Superior is the largest collection of fresh water in the known world. It is 490 miles long, and about 1700 miles in circumference.

Q. Which are the principal *rivers?*

A. Macken'zie's, Nelson's, St. Lawrence, Missssip'pi, Missouri, (Mis-soo'-ree) Ri'o del Norte, Colora'do, and Colum'bia, or Or'egon.

1. *Macken'zie's river, 1st class, and Nelson's river, 2nd class,* both rise in the Rocky mountains; the former *flows* northerly into the Frozen Ocean; the latter *flows* easterly into Hudson's Bay.

2. The *St. Lawrence, 1st class,* whose *general course* is from S. W. to N. E. is the outlet of the great western lakes. It is navigable for ships of the line to Quebec, and for large ships to Montreal, 580 miles from the sea. It *flows* into the gulf of St. Lawrence.

3. The *Mississippi, 1st class, rises* near the west end of Lake Superior, and *running* southerly *flows* into the gulf of Mexico. It is *navigable* for boats of 40 tons 2400 miles; and for ships to Natchez, 400 miles.

4. The *Missou'ri, 1st class,* is the western branch of the Mississippi. It *rises* in the Rocky mountains, *flows* southeast, and is *navigable* for boats to the Great Falls, 3970 miles from the Gulf of Mexico.

5. The *Rio del Norte, 1st class, Colora'do, and Oregon, 2nd class,* all *rise* in the Rocky mountains; the *first flows* south

easterly into the gulf of Mex'ico; the *second flows* southwesterly into the gulf of Califor'nia; and the Or'egon, which signifies "the river that flows to the *west*," into the Pacific.

Q. Which are the principal *straits*?

A. Baffin's, or Davis' strait, Hudson's strait, strait of Belle Isle, (Bel-ile') and Behring's strait.

Q. Which are the most noted *capes*?

A. Capes Farewell, Sable, Cod, Hat'teras, Lookout, Fear, and St. Lu'cas.

Q. What is the *gulf stream*?

A. It is a current of water, which proceeds from the gulf of Mexico, along the eastern coast of North America to Newfoundland', where it meets another current from Baffin's Bay, both of which are lost in the Atlantic Ocean.

Q. What is the *character* of the North American Indians?

A. They are intelligent, grave, courageous, and warlike; but cruel and revengeful.

Q. Which are the *divisions* of North America?

A. The United States, Mex'ico, Guatemala, (Gwah-te-mah'-la) British America, the Russian settlements on the North West Coast, and Green'land, which belongs to Denmark.

Questions.—Map of N. America.

What oceans lie north, east, and west of North America? In what part are the British possessions? the United States? the Russian possessions? Greenland? Which way from the United States is Mexico? Which way from Mexico is Guatemala? What is the most eastern point of North America? What is the most western? What straits separate it from Greenland? What from Asia? What isthmus separates it from S. America? *Between what latitudes does N. America lie?*

In what direction does the western coast of North America run? In what direction the eastern? In what direction the Rocky Mountains? Which are the five largest bays or gulfs in North America? What large islands lie near the mouth of the St. Lawrence? What islands east of Carolina? Which are the five largest lakes in North America? Is California on the eastern or western side of North America? What strait separates America from Asia? Which is the principal river that empties into Hudson's bay? Into the gulf of St. Lawrence? Into the gulf of Mexico? Into the gulf of California? Into the Pacific Ocean?

*Where do the following rivers rise, and which way do they flow?
viz. Mackenzie's and Nelson's rivers? St. Lawrence? Mississippi? Missouri? Rio del Norte? Colorado? Columbia, or Oregon?*
What separates Labrador from Greenland? Labrador from Newfoundland? Where are capes Farewell? Sable? Cod? Hatteras?
Lookout? Fear? St. Lucas? *In what direction from Halifax is
Boston? In what direction from Halifax is Quebec? In what direction is New York from Montreal? In what direction from New
York is New Orleans? Which way from New Orleans is Mexico?*

United States.

United States Capitol, Washington.

Q. How are the United States *bounded?*

A. N. by B... A...;   E. by B... A..., and the A...;
S. by the G... of M..., and M...; W. by M..., and the
P... O...

1. The *United States* are about 3000 miles from *east* to *west*,
and about 1700 from *north* to *south;* containing 2,000,000 of
square miles, and a *population* of about 12,000,000, of which
more than one million and a half are slaves.

2. The *eastern states* are generally uneven. *South of Long
Island, the coast,* in some parts, for more than 100 miles from
the sea, is mostly a flat, sandy plain. *West* of this tract, the
country is hilly. The country *between* the Al'leghany and
Rocky mountains is moderately uneven.

3. The United States present nearly every variety of *climate.* The *spring* season of the southern extremity of the
union, is generally about two and a half months in advance of
that of the northern extremity. *More rain falls* in the United
States annually than in Europe.

Q. What is the *capital* of the United States ?

A. Wash'ington, in the District of Colum'bia.

Washington, 6th class, is situated on the N. E. bank of the Potomac, 300 miles from the mouth of the river. It is regularly laid out, and should it ever be completed, according to the original plan, it will be one of the most magnificent cities in the world. *The capitol,* in which Congress meets, is a splendid edifice. It consists of a central edifice, and two wings. The wings are 100 feet square, and the whole building presents a front of about 362 feet.

Q. Which are the principal *mountains?*

A. The Al'leghany and Rocky mountains.

Q. Which are some of the principal *rivers* ?

A. The Connect'icut, Hud'son, Del'aware, Poto'mac, Savan'nah, Ohi'o, Tennessee', Mississip'pi, Missou'ri, Arkansas', (Arkansaw') Red, and Or'egon, or Colum'bia.

Q. Which are the principal *productions ?*

A. Grass is the principal production of the *eastern* states ; wheat and tobacco of the *middle;* cotton, rice, and sugar, of the *southern,* and wheat, Indian corn, hemp, cotton, and tobacco, of the *western.*

Q. Which are the principal *exports?*

A. Cotton, flour, tobacco, timber, and rice.

Q. *When* and *by whom* were the United States first *settled ?*

A. The first permanent settlement was made by the English, at James'town, Virgin'ia, 1607 ; the first settlement in New England, was at Ply'mouth, Massachu'setts, 1620.

Q. To whom did the United States formerly *belong* ?

A. They were colonies of Great Britain, until July 4th, 1776, when, consisting of thirteen states, they declared themselves free and independent.

1. *Hostilities were commenced* between the colonies and Great Britain, in 1775. The *first blood was shed* at Lex'ington, Massachusetts.

2. The *independence* of the United States was *acknowledged by Great Britain,* in 1783.

3. The 13 *states,* which at the time of the declaration of independence *composed the confederacy,* were New Hampshire

Massachu'setts, Rhode Isl'and, Connect'icut, New York, New Jersey, Pennsylva'nia, Dela'ware, Ma'ryland, Virgin'ia, North Caroli'na, South Caroli'na, and Geor'gia.

4. The *present constitution* of the United States was *adopted* in 1788.

Q. What is the *government*?

A. It is a federal republic, being an union, or confederation of several states, each a republic, under the general power of a congress.

1. The *constitution secures* to the citizens the grand principles of freedom, liberty of conscience in matters of religion, liberty of the press, trial by jury, &c. &c.

2. The *executive power*, that is, the power which *administers* the government, is *committed* to a president who is chosen once in four years, by electors appointed by the several states.

3. The *legislative power*, that is, the power which *enacts* all the laws, is *vested in* a congress, consisting of a senate and house of representatives.

4. The *senate consists* of two members from each state, chosen by the legislature for six years.

5. The *representatives* are elected by the people every two years. One representative is chosen for every 40,000 inhabitants. In the slave holding states, 5 slaves are allowed to count the same as 3 freemen.

6. The *judiciary*, which *expounds and applies* the laws, is independent of the legislature. The judges hold their office during good behaviour.

7. *Each of the states is an independent republic*, and has a separate executive, legislature, and judiciary, with a constitution of government similar to that of the United States.

Q. How *many* states are there in the United States, and how may they be *divided*?

A. There are 24 states, and 6 territories; which may be divided into *eastern, middle, southern,* and *western* states.

Q. Which are the *eastern states*?

A. Maine, New Hampshire, Vermont, Massachusetts, Rhode Island, and Connecticut, which are commonly called the New England states.

Q. Which are the *middle states*?

A. New York, New Jersey, Pennsylvania, and Delaware.

Q. Which are the *southern states?*

A. Maryland and Virginia, between which is the District of Colum'bia; also North Caroli'na, South Caroli'na, Geor'gia, Alaba'ma, Mississip'pi, and Louisia'na.

Q. Which are the *western states?*

A. Tennessee', Kentuc'ky, Ohi'o, Indiana, (Inje-an'na) Illinois, (Il-le-noy') and Missouri, (Missoo're.)

Q. Which are the *territories?*

A. Michigan, (Mish'e-gan) North West, Missouri, Western, Arkansas, (Ar-kan-saw') and Flor'ida.

Questions.—*Map of the United States*

What lakes lie on the N. of the United States? What ocean on the E.? What Gulf on the S.? What bay is there on the coast of Mass.; and what are its capes? What islands S. of Mass. and Rhode Island? What island and sound S. of Connecticut? What is the eastern point of Long Island called? What bay lies S. of Pennsylvania? What are its capes? What bay and capes S. of Maryland? Which bay is the largest? What sounds are there on the coast of N. Carolina? What inlets lead into them? What capes on this coast?——What is the principal range of mountains in the U. S. east of the Mississippi river? Through what states do they pass? What branch is in Tennessee? Where do the rivers empty which rise on the eastern side of these mountains? What great river receives most of those on the western side? What ranges of mountains are in the Northeastern states?——Where is the Hudson? What is its principal branch? *Where does the Hudson rise? What is its course? Where does it empty?* Which are the principal rivers emptying into the Atlantic east of the Hudson? Which is the largest? *Where do these rivers rise? Where do they empty?* What rivers empty into Albemarle sound? What into Pamlico sound? What large rivers empty into the Gulf of Mexico? What is the general course of rivers which empty into the Atlantic S. of the Susquehannah? What rivers empty into Chesapeak Bay? What rivers are between Pamlico sound and Savannah river? *Describe Savannah river, i. e. tell where it rises, what is its course, where it empties.* What is the principal river of East Florida? Mention the Eastern and Western branches of the Mobile river which empties at Mobile. What two rivers between Mobile river and the Mississippi? *What is the source, course, and place of discharge of the Mississippi?* What are the great branches of the Mississippi south of the Ohio on the western side? Which are the principal on the eastern side? What are the branches which form the Ohio? *Where do they rise? Where do they unite?* What are the branches of the Ohio on the north side? On the south side? What is the

great eastern branch of the Mississippi north of the Ohio?——
Where is lake Michigan, and with what lake does it communicate?
What lake between Huron and Erie? What river empties into it?
What rivers empty into lake Erie, and from what state? What falls
between Erie and Ontario? What river empties into lake Ontario?
What lake lies east of lake Ontario? How is it connected with the
river St Lawrence? What small lake lies south of lake Champlain?

EASTERN STATES; OR NEW ENGLAND.

Q. Which are the *Eastern*, or *New England States?*

A. Maine, New Hamp'shire, Vermont', Massachu'setts, Rhode Isl'and, and Connect'icut.

1. The *western* part of New England is mountainous; the
rest is hilly, with occasional plains.

2. The *soil* is in general good, and is better adapted to graz-
ing than to tillage.

3. The *climate* is subject to great extremes of heat and
cold, but is healthful. Easterly winds prevail in the spring
on the eastern coast, which are damp and extremely depress-
ing.

Q. Which are the principal *mountains?*

A. The *Green* mountain, and the *White* mountain ranges, which run from north to south, through
the whole length of New England.

1. The *Green* mountain range *commences* near the Canada
line, and *terminates* at New Haven, in the southern part of
Connecticut; the *highest peak is called* Killington, in Vermont, 6th class.

2. The *White* mountain range, *highest peak* in N. H. 5th
class, also *commences* near the Can'ada line; *divides* below
Northamp'ton, Mass. into two branches, one of which, called
Mt. Tom range, *terminates* at New Ha ven; and the *other* at
Lyme, near the mouth of the Connecticut river.

Q. Which is the principal *river* in New England?

A. The Connecticut.

1. The *Connecticut*, 4th class, *rises* near the Canada line,
and *flows* southerly; it separates Vermont from New Hampshire, and passing through Massachusetts and Connecticut,
flows into Long Island sound. It is *navigable* for sloops to
Hartford, 50 miles, and for boats 300.

2 The *other principal rivers* are the Penob'scot, 5th class; Kennebock', Mer'rimack, and Housaton'ick, 6th class.

Q. Which are the *chief towns?*

A. *P...d* is the chief town of Maine; *P...th* of New Hampshire; *B...n* of Vermont; *B...n* of Massachusetts; *P...ce* of Rhode Island; and *N.w H...n* of Connecticut.

Q. Which are the *capitals* or *seats of government* of these states?

A. *P...d* is the capital of Maine; *C...d* of New Hampshire; *M...r* of Vermont; *B...n* of Massachusetts; *P...ce* of Rhode Island; and *H...d* and *N.w H...n* of Connecticut.

By the *chief town* of a state, is meant that whose inhabitants are most *numerous;* the *capital* denotes the *seat of government.*

Q. Which is the principal *religious denomination?*

A. The Congregational denomination; next to which the Baptists are the most numerous, after whom are Episcopalians and Methodists.

Q. Which are the *colleges* of New England?

A. *Bruns'wick* college in Maine; *Han'over* in New Hampshire; *Middlebu'ry* and *Bur'lington* in Vermont; *Harvard* University, *Wil'liamstown* and *Amherst* colleges in Massachusetts; *Providence* in Rhode Island, and *Yale and Washington* colleges in Connecticut.

Q. What can you say of New England as to commerce?

A. It is the most commercial division of the United States.

Q. Which is the most important *mineral?*

A. Iron, which is found in many places.

Q. Which are its principal *manufactures?*

A. Cotton and woollen goods, hats, shoes, iron, and tin ware.

Q. Which are its principal *productions?*

A. Beef, pork, butter, cheese, and grain.

Q. Which are its principal *exports?*

A. Beef, pork, butter, cheese, grain, timber, pot and pearl ashes, and fish.

Q. What is the *character* of the people of New England ?

A. They are intelligent, enterprising, industrious, brave, and moral; but very fond of gain.

Questions.—*Map of the United States.*

Which are the eastern states? which is the largest? which the smallest? In what part of New Hampshire are the White Mountains? Where are the Green Mountains? Where is Passamaquoddy Bay? Massachusetts Bay? Machias, Penobscot, Casco Bay? Nantucket Island? Long Island? Block Island? What celebrated point is at the east end of Long Island? Where is Cape Ann? Cape Cod? Cape Malabar? *What is the course and place of discharge of the Connecticut? Merrimack? Kennebeck? Penobscot?* Point out the chief towns of the several eastern states, *and tell where situated;* also the capitals. *What is the latitude and longitude of Boston?* Which way from Boston is Hartford? Portland? Concord?

MAINE.

Q. How is Maine *bounded?*

A. N. by L... C...; E. by N... B...; S. by the A...; and W. by N... H..., and L... C...

1. Scattered *settlements* began to be made in Maine, as early as 1630. In 1651, or 1652, the province was attached to Massachusetts, under whose jurisdiction it continued, until 1820, when it became an independent state.

2. Along the *sea coast*, the land is poor: but a rich soil exists between the Penob'scot and Kennebeck' rivers. A large portion of the northern part of the state is still a wilderness.

3. It has a *coast* of more than 200 miles, abounding with fine harbours.

4. *Greatest length,* 225 *miles; greatest breadth,* 195; *square miles,* 32,000. *Pop.* 298,335.

Q. Which are its principal *bays?*

A. Passamaquod'dy, Machi'as, Casco, and Penob'scot.

Q. Which are its principal *rivers?*

A. Penob'scot, Kennebeck', Androscoggin, Saco, and Piscat'aqua.

The *Penobscot,* 5th class, rises in the north eastern part of the state, and *flows* east and south into Penobscot bay. The *Kennebeck,* 6th class, rises near the same place, and *flows* southerly into the Atlantic. The *Androscoggin,* 6th class, is a branch of the Kennebeck, and joins it near its mouth. The *Saco,* 6th class, rises among the White mountains in New Hampshire, and *flowing* south east, empties itself west of Portland. The *Piscataqua, 6th class, passes into* New Hampshire, and *empties* itself at Portsmouth.

Q. Which is the *chief town* and *capital?*

A. P...d.

1. *P...d, large town,* is built on a peninsula, in Casco bay. It has a safe and capacious harbour.

2. The *other principal towns* are Eastport, Penobscot, Hallowell, (Hal'lo-ell) Machias, Bath, Wiscas'set, and York, *all small towns.*

Q. What *college* does this state contain?

A. Bowdoin college at Brunswick.

1. This college has a large property in lands. Its librar-contains about 5000 volumes.

2. Besides this *institution,* a Congregational Theological se minary, has been established at Ban'gor; and a Baptist Theological seminary, at Waterville'.

Q. Which are the principal *islands* on the coast of Maine?

A. Mount Desert' Island, Long Island, Deer Island, Seguin Island.

Mount Desert is 15 miles long by 12 broad, and contains about 1200 inhabitants. On Se'guin Island, at the mouth of Kennebeck river, is a light house, with a repeating light.

Q. Which are the principal *religious denominations?*

A. Congregationalists and Baptists.

Q. Which are the principal *productions?*

A. Grain and grass.

Q. Which are the principal *exports?*

A. Lumber and fish.

Questions.—*Map of the United States.*

How is Maine bounded on the N. E. S. W.? What are its principal rivers? *Where do they rise? What is their direction? Where do they empty?* Mention the principal bays. What is the capital? *Where situated? Where is Eastport?* What island lies off the coast?

NEW HAMPSHIRE.

Q. How is New Hampshire *bounded?*

A. N. by L... c...; E. by M..., and the A...; S. by M...; and W. by v... and L... c...

1. The first *settlements* in New Hampshire, *were begun in* 1623, by the English, on the river Piscataqua.

2. It has only 18 miles of *sea-coast,* and Portsmouth is the only sea-port.

3. The land on the *sea-coast* is level; in the *interior* it

rises into hills and mountains. The *soil* is difficult of cultivation.

4. The *climate* is cold ; but healthful.

5. Its *length* is 100 miles ; *breadth* at the southern extremity, 90 ; *square miles*, 9500. *Pop.* 244,161.

Q Which are the principal *mountains?*

A. The White mountains.

This is the highest land in the United States, east of the Mississippi. The loftiest peak is Mount Wash'ington, *5th class* The Grand Monad'nock, *6th class*, is in the southern part o. the state.

Q. Which are the principal *rivers ?*

A. The Connect'icut, Mer'rimack, Piscat'aqua.

The *Connecticut, 4th class*, divides this state from Vermon. The *Merrimack, 5th class, rises* in the White mountains, ar *running* southerly, passes into Massachusetts. The *Piscataqua, 6th class, empties* into the ocean at Portsmouth.

Q. Which is the principal *lake ?*

A. Win'nipiseog'ee (Win'ne-pe-sog'e) lake.

This lake is near the centre of the state, and is 23 miles long.

Q. Which is the *chief town*, and which the *capital ?*

A. P...th is the chief town, and C...d the capi tal.

.1. *P...th, large town,* is *situated* at the mouth of the Piscataqua, and has one of the best harbours cn the continent.

2. *C...d, small town, lies* on the Merrimack, by means of which, and a canal, it communicates with Boston.

3. The other *principal towns* are Ex'eter, Charles'town, Ha'-verhill, and Keene, *small towns.*

Q. What *college* does this state contain ?

A. Dartmouth college, at Han'over.

1. *Dartmouth college* is one of the oldest colleges in the United States. A medical school is connected with it.

2. An academy called Philip's academy, in Exeter. is a flourishing institution. It has a library of 700 volumes, and funds which amount to 80,000 dollars.

Q. Which are the principal *religious denominations ?*

A. Baptists and Congregationalists.

Q. Which are the principal *productions ?*

A. Grass and grain.

Q. Which are the principal *exports ?*

A. Lumber, iron, and the products of the dairy.

Questions.—*Map of the United States.*

How is New Hampshire bounded N. E. S. W. ? What range of mountains does the state contain ? What river rises in this state, and passes into Mass. ? What lake can you mention ? What is the chief town ? What the capital ? *How are these towns situated? What sea-port has N. H. ? What is the extent of sea-coast ?*

VERMONT

Q. How is Vermont' *bounded?*

A. N. by L... C... ; E. by N.... H....; S. by M.... . W. by N... Y...

1. *Vermont* was originally a part of New Hampshire, and was *settled* at a much later period than any of the eastern states. *It was admitted into the union* in 1791.

2. The *face of the country* is mountainous.

3. Most of the state has a good *soil,* and a healthful *climate.*

4. Its *length* is 157 miles ; *breadth* on the northern border, 90 miles ; and 40 on the southern; *square miles* 10,200. *Pop.* 235,764.

Q. Which are the principal *mountains?*

A. The Green mountains.

The *Green mountains,* so called from their perpetual verdure, run from north to south through the state, and pass into Massachusetts. The *highest peaks* are Killington, Camel's Rump, and Mansfield, 6th *class.*

Q. Which is the principal *lake?*

A. Lake Champlain' (Shamplain').

1. *Champlain'* is 100 miles long, and from one to five miles broad. It *communicates with the* St. *Law'rence* by the river Sorelle'.

2. *Memphrema'geg* is a smaller lake, lying partly in this state, but principally in Lower Canada.

Q. Which are the principal *rivers?*

A. The Connect'icut, On'ion, Ot'ter Creek, La-moile', and Missis'que.

The *Connecticut,* whose *course* is southerly, forms the boundary between Vermont and New Hampshire. The *other rivers* which are small, *generally flow* westerly, and *fall* into lake Champlain.

Q. Which is the *chief town,* and which the *capital?*

A. B...n is the chief town, and M...r is the capital.

1. *B....n, small town,* lies in the southwest corner of the state. It is celebrated for the defeat of a party of Hessians from Burgoyne's army by Gen. Stark, in 1777.

2. *M...r, small town, lies* on Onion river.

3. The *other principal towns* are Windsor, Rutland, Middlebury, and Burlington, *small towns.*

Q. What *colleges* does this state contain?

A. The university of Vermont, at Burlington, and a college at Middlebury.

Q. Which are the principal *religious denominations?*

A. Baptists and Congregationalists.

Q. Which are the principal *productions?*

A. Grass, grain, and cattle.

Q. Which are the principal *exports?*

A. Live cattle, beef, pork, pot and pearl ashes.

Questions.—Map of the United States.

How is Vermont bounded N. E. S. W.? What range of mountains does the state contain? What lake lies on the west? By what does it communicate with the St. Lawrence? What lake lies in the north? Which is the chief town? Which the capital? *How are they situated?*

MASSACHUSETTS.

Q. How is Massachusetts *bounded?*

A. N. by v..., and n... h....; E. by the a...; S. by the a..., r... i..., and c...; and W. by n... y...

1. Massachusetts was the first *settled* state in New England, the colonists, 101 in number, landing at Plymouth, Dec. 22, 1620. The anniversary is still observed.

2. The *surface of the state* is greatly diversified; *near the coast,* the land is level; the *interior* is hilly, and the *western part* mountainous.

3. The *climate* is dry and healthful, except on the coast, where easterly winds render the atmosphere much of the time damp and unpleasant. The *soil* is productive, and agriculture is in a high state of cultivation.

4. The *length* of Massachusetts, on the northern line, is 130 miles; *breadth* at the western extremity, 50 miles; square miles, 7250. Pop. 523,287.

Q. Which are its principal *mountains?*

A. The range of *Green* Mountains extends from N. to S. through the western part of this state.

The most *noted* mountain in Massachusetts is Wachusett, in Princeton, *6th class.*

Q. Which are the principal *bays?*

A. Massachusetts bay, between Cape Cod and Cape Ann, and Buzzard's bay, on the southwest side of Cape Cod.

Q. Which are the most noted *capes?*

A. Ann, Cod, and Malabar.

Q. Which are the principal *islands?*

A. Nantuck'et, Martha's Vineyard, and Elizabeth Islands.

1. *Nantucket Island* is 15 miles *long.* It *contains* about 7000 *inhabitants*, who are distinguished for their skill and enterprise, in the whale fishery. They hold all their lands in common. All their cows, amounting to 500, feed together in one herd, and all their sheep, 14,000, in one pasture.

2. *Martha's Vineyard* is about 20 miles *long.* The *Elizabeth Islands* are several in number, and lie in a row of about 18 miles in length.

Q. Which are the principal *rivers?*

A. Connect'icut and Mer'rimack.

The *Connecticut, 4th class,* only passes through this state in a southerly *direction;* the *Merrimack, 5th class,* comes from New Hampshire, and *running* northeasterly, *empties* itself into the ocean at Newburyport.

Q. Which is the chief *town* and *capital?*

A. B...n.

1. *B...n, 4th class,* is *situated* at the head of. Massachusetts bay. It is the chief city of New England, in commerce, population, and wealth. It is the second commercial city in the United States. Its literary and humane institutions are numerous and distinguished. Its wealth is unusually great. The scenery around Boston is scarcely excelled in any country.

2. The *other principal towns,* are, Salem, *6th class,* Charles'town, *large town,* north of Boston, celebrated for the battle of Bunker's Hill, June 17, 1775; Newburyport', *large town,* and Lynn, Spring'field, Worcester, (Woors'ter) New Bedford, Northamp'ton, *small towns.*

Q. What *colleges* does this state contain ?

A. The University at Cambridge, near Boston; Williams' College, at Williamstown, and Amherst College, near Northampton.

1. The *University at Cambridge,* is the most ancient and wealthy literary institution in the United States. It has 20 professors, and a library of 26,000 volumes. The other co leges are respectable, and liberally endowed.

2. Besides these, there is a *Theological seminary* of great ce lebrity at Andover, 20 miles north of Boston, and several *respect able academies,* the principal of which is Phillips' academy, at Andover; Dummer Academy, at New'bury, and Leicester (Les'ter) academy, at Leicester.

Q. Which are the principal *religious denominations ?*

A. Congregationalists and Baptists

Q. Which are the principal *productions ?*

A. Grass and grain.

Q. What can you say of the *fisheries ?*

A. Very extensive business is done, both in the cod and whale fisheries.

Q. What of the *manufactures?*

A. Manufacturing establishments are numerous and increasing. The principal manufactures are cotton goods, shoes, hats, and glass.

Q. What of *commerce?*

A. The commerce of Massachusetts exceeds that of any other state in the Union, except New York; it exports fish, pot and pearl ashes, beef, pork, &c. &c.

Questions.—Map of the United States.

How is Massachusetts bounded N. E. S. W. ? Which are the principal mountains? What bay lies in the east ? What are its capes ?—What cape lies S. of Cape Cod ? What islands lie S. of Cape Malabar ? What river comes from N. Hampshire ? What river passes through this state ? Which side of Connecticut river is there most land in the state ? What canal do you notice, and what does it connect ?—Which is the chief town and capital ? *How situated ?*

RHODE ISLAND.

Q. How is Rhode Island *bounded?*

A. N. and E. by м....; S. by the л....; and **W.** by с....

1. Rhode Island was first *settled* in 1636, by Roger Williams, who was banished from Massachusetts on account of novel opinions; it has never formed a constitution, but is still governed by the charter it received from King Charles II.

2. The *southern* part is level, sandy, and barren; the *northern* is hilly. The islands, and some portions of the coast, are remarkably beautiful, healthful, and fertile.

3. *Length*, 50 miles; *breadth*, on its northern border, 29 miles; *square miles*, 1500. *Pop.* 83,059.

Q. What *bay* does this state contain?

A. Narragan'sett bay, which divides the state into two parts.

Narragansett bay is about 28 miles *long*, and 10 *broad*.—The *northeastern extremity is called* Mount Hope bay; the *northwestern*, Greenwich bay, and the *southern*, Providence bay.

Q. Which are the principal *rivers?*

A. Pawtucket, Providence, and Pawtuxet.

Pawtucket river, *6th class, rises* in Worcester county, in Massachusetts, and *flowing* southeasterly, *falls* into Providence river, one mile below the town of Providence. *Providence* river, *6th class*, is formed by two small rivers which unite just above Providence. *Pawtuxet* river, *6th·class, falls* into Providence river 5 miles below the town of Providence.—This last river abounds with falls, which furnish numerous fine mill seats.

Q. Which are the principal *islands?*

A. Rhode Island and Block Island.

Rhode Island, on account of its salubrity, fertility, and beauty, has been styled the *"Eden of America."*

Q. Which is the *chief town* and *capital?*

A. P...ce.

1. *P...ce, 6th class, lies* on the river of that name, about a mile above the mouth of the Pawtucket, and 35 miles from the ocean. It is the third town in New England in point of population. Its commerce is extensive.

2. The *other principal towns* are New'port, *large town*, distinguished for its fine harbour and salubrious situation; Bristol and War'ren, *small towns*.

Q. What *Indians* live in this state?

A. The remains of the once famous Narragan'sett tribe, now reduced to about 100 souls.

1. The *Narragan'setts* were one of the tribes engaged in the Indian war against the colonies in 1675, called King Philip's war, which for a time threatened their existence.

2. In the course of the war, the fort of the tribe was taken, after a terrible contest, and their village, consisting of 600 wigwams, was burnt by troops from Massachusetts and Connecticut. Seven hundred Indian warriors were killed on the spot, and 300 died of their wounds: 300 were taken prisoners, and as many women and children. From this defeat the tribe never recovered.

Q. What *college* does the state contain?

A. Brown University, in Providence, which is a respectable institution.

Learning is not so generally diffused among the inhabitants of Rhode Island, as among those of the other New England States. Common schools are not supported by law.

Q. Which is the principal *religious denomination?*

A. The Baptists.

Q. What can you say of its *manufactures?*

A. According to its population, it is the most considerable manufacturing state in the Union. The principal article is *cotton goods.*

Q. Which are the principal *productions?*

A. Beef, pork, butter, and cheese.

Q. Which are its principal *exports?*

A. Cotton and linen goods; hats are also exported in considerable quantities.

Questions.—*Map of the United States.*

How is Rhode Island bounded N. E. S. W.? What island is S. of Rhode Island? Which is the chief town and capital? *How situated?*—Which way from Boston is Providence? Which way from Providence is Newport?

CONNECTICUT.

Q. How is Connecticut *bounded?*

A. N. by M...; E. by R... I...; S. by L.... I.... s..., and W. by N... Y...

1. The *first house erected* in Connecticut was at Windsor, in 1633, by some of the Plymouth colonists. In 1635, *Windsor, Wethersfield*, and *Hartford* were *settled* by about 60 men, women, and children, from Massachu'setts. The state originally *contained* two *colonies*, Connect'icut and New Ha'ven, which were united in 1665. The *present constitution was formed* in 1818.

2. The *surface* of Connecticut is generally uneven, rising into mountains in the northwestern parts. The *soil* is various; that of *Connecticut valley* is uncommonly rich and fertile; the *sea-shore* is sandy and barren.

3. The *climate* on the *coast* is variable and moist; that of the *interior* is more healthful and agreeable.

4. *Length*, 72 miles; *breadth*, on the eastern boundary, 45; *square miles*, 4700. *Pop.* 275,248.

Q. What are the principal *rivers* ?

A. Connect'icut and Housaton'ick, or Stratford.

The Connecticut, *4th class, comes* from Massachusetts, and *flowing* south, *empties* itself into Long Island sound, at Saybrook. The *Housatonick, 6th class, rises* in Massachusetts, and *flows* southeast into Long Island sound, west of New Haven.

Q Which is the *chief town*, and which are the *capitals?*

A. N...w H...n is the chief town, and H...d and N...w H...n the capitals.

1. *N..w H...n, large town*, is a beautiful town, situated on a bay, which sets up from Long Island sound.

2. The legislature *meets* alternately at Hartford and New Ha'ven. *H.....d*, also a *large town*, is on the Connecticut river, 50 miles from its mouth.

3. The *other principal towns* are New Lon'don, Norwich, (Nor'rij) Middletown, Wethersfield, Litchfield, and Saybrook, *small towns*.

Q. What *colleges* does this state contain ?

A. Yale College at New Haven, and Washington College at Hartford.

1. *Yale College was founded* in 1701. Its library contains about 8000 volumes. The chemical and philosophical apparatus are handsome and complete. It contains the noblest collection of minerals in the United States, which has recently been purchased of Col. Gibbs, for the sum of 20,000 dollars.

4*

2. *Washington College* has recently been incorporated.

3. *In Hartford there is an interesting institution* called the American Asylum for the education of the Deaf and Dumb, and an Insane Hospital. At *Litchfield* there is a respectable Law *School.*

4. This state has a *fund* of nearly two millions of dollars, the interest of which is appropriated to the support of common schools.

Q. Which are the principal *religious denominations?*

A. The Congregational, Episcopalian, and Baptist.

Q. Which are its principal *manufactures?*

A. Tin, fire arms, nails, hats, and cotton goods.

Q. Which are its principal *exports?*

A. Beef, pork, Indian corn, butter, and cheese.

Questions.—Map of the United States.

How is Connecticut bounded N. E. S. W. ? Which is its principal river ? *Where does it empty?* Which is its chief town ?—Which its capitals ? *How are these places situated?* What island lies S. of Connecticut ?

MIDDLE STATES.

Q. Which are the *Middle States?*

A. New York, New Jer'sey, Pennsylva'nia, and Del'aware.

1. The *northern and western parts* are hilly, and, in some places, mountainous. The *southern parts* are generally level

2. The *soil* is various, but generally good ; in some places it is very fertile.

3. The *climate* is healthful. The weather is more liable to frequent and sudden changes ; but the winters are milder than those of New England.

Q. Which are the principal *rivers?*

A. The Hud'son, Del'aware, and Susquehan'nah.

1. The *Hudson, 4th class, rises* west of lake Champlain, and *flows* southerly into the Atlantic. It affords excellent *navigation* for large vessels to Hudson, 130 miles, and for sloops to Troy, 166 miles.

2. The *Delaware, 5th class, rises* in the Catskill mountains in New York, and *running* a zigzag course, separates New York and New Jersey from Pennsylvania, and *flows* into Delaware bay. It is *navigable* for 74 gun-ships to Philadelphia, 40 miles.

3. The *Susquehannah, 4th class,* the largest river in Pennsylvania, is *formed* of two branches; the *eastern branch rises* in New York, and the *western* in Pennsylvania. They *unite* at Northumberland, and the river *flows* first south and then southeast into Ches'apeak bay. Its *navigation* is much obstructed by falls and rapids.

Q. Which are the *chief towns?*

A. *N...w Y...k* is the chief town of New York; *N...w B...k* of New Jersey; *P...a* of Pennsylvania; and *W...n* of Delaware.

Q. Which are the *capitals* of these states?

A. *A...y* is the capital of New York; *T...n* of New Jersey; *H...g* of Pennsylvania; and *D...r* of Delaware.

Q. What *colleges* do the middle states contain?

A. *Colum'bia, Un'ion,* and *Ham'ilton* colleges in New York; the *College of New Jersey* in New Jersey, and the *University of Pennsylvania, Dick'inson, Jefferson, Al'leghany,* and *Wash'ington* colleges in Pennsylvania.

Q. Which is the principal *religious denomination?*

A. Presbyterian; next to whom are Episcopalians and Methodists.

Q. What can you say of the middle states as to *commerce?*

A. The commerce of these states centres chiefly in the cities of New York and Philadelphia.

Q Which is the principal *production?*

A. Wheat: other kinds of grain, however, are extensively cultivated.

Q. What is the *character* of the people of the middle states?

A. In the higher classes, they are more refined than those of the eastern states; but the lower classes are more rude, ignorant, and vicious.

Questions.—Map of the United States.

Which are the middle states? Which is the largest? Which next? Which the smallest? In what direction do the Alleghany mountains run? Where are the Catskill mountains? Where is Delaware Bay? Between what capes does it open into the ocean? Which way from lake Ontario is lake Erie? What connects these lakes? What falls are between them? What states does lake Champlain separate? *What is the general course and place of discharge of the Hudson? Mohawk? Genesee? Delaware? Susquehannah rivers? Where are the chief towns situated? Where the capitals?*

NEW YORK.

Q. How is New York *bounded?*

A. N. by u... c...., and l... c....; E. by v..., m..., and c...; S. by the a..., n... j..., and p...; and W. by p..., and u... c..., from which it is separated by l... e...., and l... o...

1. The *state of New York was first settled* in 1614, by some Dutch adventurers, who built a fort at Albany, on Hudson river. The next year, a settlement was begun where New York now stands. The Dutch *held possession of the territory* till 1664, when it was taken by the English.

2. The *northern parts* of New York are rugged and barren. The *eastern* and *western* parts are rich and fertile.

3. The *climate* of the *northeastern* part is extremely cold; the *western* is milder than the eastern, but is less healthful.

4. *Length,* 340 miles; *breadth,* 300; *square miles,* 46,000. Pop. in 1826, 1,616,458.

Q. What *bay* has this state?

A. New York harbour is a bay, which extends 9 miles south of the city, and is from 1 to 5 miles broad.

Q. Which are the principal *mountains?*

A. The Catskill mountains.

The *Highest Peak* in this range, is Round Top, 6th *class.*

Q. Which are the principal *rivers?*

A. The Hud'son, Mo'hawk, and Genesee'.

The *Hudson,* 4th *class, rises* in the northern part of the state, and *flows* southerly into the Atlantic, below New York. The *Mohawk,* 6th *class,* whose *course* is southeast, is a branch of the Hudson, with which it *unites* between Waterford and Troy. The *Genesee, sixth class, rises* in Pennsylvania, and *flows* northerly into lake Onta'rio, below Roch'ester.

Q. Which are the principal *islands?*

A. Long Island, Stat'en Island, and Manhat'tan Island.

Long Island extends east of New York 140 miles; it is, on an average, about 10 miles *broad. Staten Island* is 18 miles *long.*

Q. Which are the principal *lakes?*

A. Onei'da, Skeneat'eles, Owas'co, Cayu'ga, Sen'eca, Crooked, and Canandai'gua Lake.

1. Cayuga and Seneca are the *largest;* but all these lakes are of moderate size.

2. Besides these lakes, the southern half of lake *Ontario,* the southeastern part of lake *Erie,* and the western part of *Champlain, belong to* New York.

Q. Which is the *chief town,* and which the *capital?*

A. N...w Y...k is the chief town, and A...y the capital.

1. *N...w Y...k, 3d class,* is the first commercial city in America. It is *situated* on the south end of Manhattan Island.

2. *A...y, 6th class,* is the second city in the state, and lies on the western bank of the Hudson, 144 miles north of New York.

3. The *other principal towns* are Troy, *large town,* Hudson, Poughkeepsie, (Po-kep'sy) New'burgh, Schenec'tady, U'tica, Gene'va, Canandaigua, Buf'falo, *small towns.*

Q. What *Indians* does this state contain?

A. Part of the remains of the Iroquois, or Six Nations, amounting to about 5000.

The *principal tribes* are the Mo'hawks, the Sen'ecas, and the Onei'das. The *Mohawks live* in Upper Canada; the *Senecas* on the rivers in the western part of the state; and the *Oneidas* near Utica, in the county of Oneida.

Q. What *colleges* does this state contain?

A. Colum'bia college, at New York; Union at Schenec'tady; and Ham'ilton, at Clin'ton.

The state possesses a *fund* of more than $1,200,000, and 80,000 acres of land, the income of which is annually appropriated to the support of common schools.

Q. Which are the principal *religious denominations?*

A. Presbyterians, Episcopalians, and Baptists.

Q. What celebrated *mineral waters* are there in this state?

A. The Sarato'ga and Balls'ton springs, about 30

miles N. of Albany; and the New Leb'anon springs, .
29 miles S. E. of Albany.

Q. What great *natural curiosity* can you mention ?

A. The falls of Niag'ara.

1. *These falls are in* Niagara river, about half way between
lake E'rie and lake Onta'rio. Here, the whole river *falls* over
a precipice, 160 feet. The roar of the waters may, at times,
be heard 40 miles. The vapours ascending from the cataract
are often seen 60 or 70 miles.

2. The *Cohoes* (Co-hose') falls, on the Mohawk, are nearly
70 *feet perpendicular.*

Q. What *canals* are there in this state ?

A. The Erie canal, which forms a communication between Lake Erie and the Hudson river; and
the Champlain canal, which unites the same river
with lake Champlain.

The *Erie canal extends* from Albany to Buffalo, a distance
of 365 *miles.* It is 40 feet wide on the surface, 28 at the bottom, and four feet deep.

Q. What article of *manufacture* demands particular notice ?

A. That of salt, at Sali'na, about 30 miles
west of Utica, where more than half a million
of bushels are manufactured annually, from salt
springs.

Q. Which are the principal *productions ?*

A. Wheat is the staple production; Indian corn.
and oats, are extensively cultivated.

Q. Which are the principal *exports ?*

A. Wheat, pot and pearl ashes, Indian corn, rye,
beef, pork, lumber, of which large quantities are
brought from the western parts of New England, and
the eastern parts of New Jersey.

Questions.—Map of the United States.

How is New York bounded N. E. S. W. ? What mountains do
you notice ? On which side the Hudson are they ? What two
large rivers rise in this state, which pass into Pennsylvania? *Where
does the Hudson rise ? What is its course ? Where does it empty ?—
Describe the Mohawk and Genesee rivers ? Trace the course of
the Grand Canal.* Is there any other canal in this state ? What
lake partly separates this state from Vermont ? What lake lies S

of Champlain ? Which is the chief town ? Which the capital ? How situated ? Which way from Albany are Saratoga Springs ? Where is Hudson ? Utica ? Rochester ? Buffalo ?

NEW JERSEY.

Q. How is New Jersey *bounded ?* -

A. N. by n... y...; E. by n... y..., and the a...; S. by the a..., and d... b...; W. by d..., and p..., from which it is separated by d... river.

1. The first *settlement* in New Jersey was *made* three or four years after the settlement of Plymouth, in New England, *by* some Dutchmen and Danes. Elizabethtown was settled in 1664. The *country was divided* into East and West Jersey, which *were united* in 1702, under the name of New Jersey.

2. The *northern part* of the state is mountainous; the *southern* is flat and sandy; the *middle* region is comparatively level, and extremely fertile.

3. The *climate* varies with the latitude and elevation. In the *north* the winters are cold, but in the *south* more temperate.

4. *Length*, 160 miles; *breadth*, 52; *square miles*, 8320.— *Pop.* 277,575.

Q. Which is the principal *bay ?*

A. Del'aware Bay.

This bay *separates* New Jersey from Delaware. Besides this may be *mentioned* Amboy bay, which lies south, and Newark bay, which lies north, of Staten Island.

Q. Which are the *capes of Delaware Bay ?*

A. Cape May, and Cape Hen'lopen.

Q. Which are the principal *rivers ?*

A. The Del'aware, Rar'itan, and Passa'ick.

The *Delaware*, 5th class *rises* in New York, and *flows* into Delaware bay. The *Raritan*, 6th class, *rises* in the eastern part of the state, and *flows* south of east into Amboy bay.— The *Passaick*, 6th class, is a small river which *flows* into Newark bay.

Q. Which is the *chief town*, and which the *capital ?*

A. N...w B...k is the chief town, T...n the capital.

1. *N...w B...k, large town, is on* the Raritan, 14 miles from its mouth, and 33 miles S. W. of New York. Steam boats from New York ascend the Raritan, as far as this place.

2. *T...n, small town, lies* on the Delaware, 30 miles above Philadelphia. The celebrated *battle* of Trenton was fought at this place, December, 1776.

3. The *other principal towns* are Newark, Prince'ton, Elizabethtown, and Bur'lington, *small towns.*

Q. What *college* does this state contain?

A. The College of New Jersey, at Princeton.

1. A *Theological* seminary was established at Princeton, by the General Assembly of the Presbyterian Church, in 1812.

2. *Queens College,* at New Brunswick, was established in 1770, but is now converted into a Theological seminary for the Dutch Reformed Church.

3. *Common Schools* are but few in New Jersey.

Q. Which is the principal *religious denomination?*

A. The Presbyterian.

Q. Which is the principal *manufacture?*

A. Iron, which is forged from *bog iron ore,* and *iron stone,* with which the state abounds.

Q. Which are the principal *productions?*

A. Wheat, rye, Indian corn; cider of excellent quality is made in this state, and many cattle are fattened for the markets of New York and Philadelphia.

Q. Which are the principal *exports?*

A. Wheat, rye, Indian corn, cattle, and cider.

Questions.—*Map of the United States.*

How is New Jersey bounded N. E. S. W.? What bay and river lie west? Which is the chief town and capital? *How situated?* Where is Cape May?

PENNSYLVANIA.

Q. How is Pennsylvania *bounded?*

A. N. by L... E..., and N... Y...; E. by N... Y..., and N...J...; S. by D..., M..., and V...; W. by V..., and O...

1. Pennsylvania was first *settled by* the Swedes, in 1627. They *held it* till 1654, when it was conquered by the Dutch; and ten years afterwards the *Dutch surrendered it* to the English. In 1681, *Charles II. granted it to* Sir William Penn, who soon after came over with a colony of Friends, and laid the foundation of Philadelphia.

2. The *middle parts* of the state are mountainous; the *eastern* and *western* parts are generally level, or moderately uneven, and fertile.

3. The *climate* in the mountainous region is cold, in winter; the *other parts* of the state are temperate, and generally healthful.

4. *Length*, 307 miles; *breadth*, 160; *square miles*, 44,000. *Pop.* 1,049,398.

Q. Which are the principal *mountains?*

A. The Al'leghany mountains, which run across the state from S. W. to N. E.

The *principal ridges* of the Alleghany mountains in Pennsylvania, are the Kittatinny, or Blue mountains.

Q. Which are the principal *rivers?*

A. The Del'aware, Susquehan'nah, and Al'leghany.

1. All these rivers *rise* in New York; the *Delaware, 5th class*, pursues a zigzag *course*, and *flows* into Delaware bay; the *Susquehannah, 4th class, passes* into Maryland; the *Alleghany, 4th class, unites* with the Monongahe'la, *5th class*, from Virginia, at Pitts'burg, and *flows* into the Ohi'o.

2. The Le'high and Schuyl'kill, both *6th class*, are *branches* of the Delaware; the former joins it at Easton; the latter at Philadelphia. The Juniat'ta, *6th class*, is a western *branch* of the Susquehan'nah.

Q. Which is the *chief town*, and which the *capital?*

A. P...a is the chief town, and H...g the capital.

1. P...a, *3d class, lies* between the Delaware and Schuylkill, five miles above their confluence. It is 110 *miles from the ocean*, by the river and bay. It is the fourth city in the Union in *amount of shipping*. In the *variety* and *extent* of its *manufactures*, it is the first city in the United States.

2. H...g, *small town, is on* the east bank of the Susquehannah, about 100 miles west of Philadelphia. It is a handsome town.

3. The *other principal towns* are Lan'caster and Pittsburg, *large towns*. Read'ing, Cham'bersburg, Car'lisle, and Wilkesbarre, (Wilks-barre') *small towns*.

Q. What *colleges* does Pennsylvania contain?

A. The *University of Pennsylvania*, at Philadelphia, to which is attached a medical school, highly celebrated through the United States; the other Colleges are *Dickinson College* at Carlisle, *Jefferson* at Can'nonsburg, *Alleghany* at Meadville, and *Washington* at Washington.

5

Q. Which are the principal *religious denominations?*

A. Presbyterians, German Calvinists, German Lutherans, Friends, or Quakers, and Baptists.

1. The *inhabitants* are of several different *nations.* One half are *English*, one fourth *German;* one eighth *Irish.* The *remainder* are.Scotch, Welch, Swedes, and Dutch.

2. The *language* generally spoken is English; but the Dutch, Germans, and Irish, *retain their own language.*

Q. What *minerals* are found here?

A. Vast quantities of coal near Pittsburg, and in other parts of the state. Iron ore abounds, and is extensively manufactured.

Q. What can you say of its *manufactures?*

A. In value and variety of manufactures, Pennsylvania is the first state in the Union. The principal manufactures are cotton goods, iron, glass, and paper.

Q. Which are its principal *productions?*

A. Wheat is the principal; next to which, is Indian corn.

Q. Which are its principal *exports?*

A. Wheat, corn, cotton goods, and glass.

Questions.—*Map of the United States.*

How is Pennsylvania bounded N. E. S. W.? What mountains pass through this state? What river is in the eastern part of the state? *Where does it rise? What is its course? Where does it empty? Describe in like manner the Susquehannah, Alleghany, and Monongahela. Where do these last rivers unite?* Which, *and where is* the capital? Chief town?

DELAWARE.

Q. How is Delaware *bounded?*

A. N. by p...; E. by d... river and bay which separate it from n... j..., and by the a...; S. and W. by m...

1. Delaware was *first settled* in 1627, *by* a number of Swedes and Fins, who, at the instance of Gusta'vus Adol'phus, king of Swe'den, emigrated to America.

2. The *northern part* of the state is hilly; the *rest is level,*

and low. The *soil* of the *southern part* is sandy and unproductive; the *soil* in the *north* is a rich clay.

3. The *climate* in the *north* is healthful and agreeable; in the *south* it is warm and moist.

4. Its *length* is 92 miles; *breadth,* 23; *square miles,* 2120 *Pop.* 72,749.

Q. Which is the principal *bay?*

A. Delaware bay.

Q. Which are the principal *rivers?*

A. Del'aware river, Bran'dywine, and Christia'na creeks.

Delaware river, 5th class, *rises* in New York, and *flows* into Delaware bay. *Brandywine,* 6th class, which *rises* in Pennsylvania, and *Christiana,* 6th class, which *rises* in Maryland, *unite* in the northern part of the state, and *empty* into Delaware river.

Q. Which is the *chief town,* and which the *capital?*

A. W...n is the chief town, and D...r is the capital.

1. *W...n, large town, lies* between Brandywine and Christiana creeks; it is *celebrated for* its flour mills.

2. *D...r, small town, is on* Jones' Creek, 7 miles above its entrance into Delaware bay, and 40 S. of Wilmington.

3. The *other principal towns* are New'castle and Lew'iston, *small towns.*

Q. Which is the principal *religious denomination?*

A. The Presbyterian.

Q. Which are its principal *manufactures?*

A. Flour is manufactured extensively on Bran'dywine creek; also gun powder, cotton and woollen goods.

Q. Which is the principal *production* and article of *export?*

A. Wheat and flour.

Questions.—Map of the United States.

How is Delaware bounded N. E. S. W.? Which is the chief town? *How situated?* Which the capital? *How situated?* What cape has the state? How does Delaware compare with other states in size?

SOUTHERN STATES.

Q. Which are the *southern states?*

A. Ma'ryland and Virgin'ia, between which is

the District of Colum'bia, North Caroli'na, South Caroli'na, Georg'ia, Alaba'ma, Missississip'pi, and Louisia'na.

1. In the southern states, the *Low Country*, that is, the tract along the sea coast, is a low, sandy plain, which varies from 50 to 200 miles in width. The *Upper*, or *Back country*, rises into hills, and at length into mountains.

2. In the *low country*, the *soil* is generally sandy, and unproductive; but on the *margin of rivers*, and in the *upper country*, it is fertile.

3. The *winters* are mild and pleasant, but in the *low country*, the *summers* are hot and sultry; and from July to the latter end of October, the *climate* is unhealthful, particularly to strangers.

Q. Which are the principal *mountains?*

A. The Al'leghany mountains, which extend through all the Atlantic southern States. They terminate in Alaba'ma.

Q. Which are the principal *rivers ?*

A. The Poto'mac, James, Roanoke', Pedee', Santee', Savan'nah, and Alatamaha, (Al-a-ta-ma-haw') all of which have a *southerly course;* and the Ap'palach'icola, Alaba'ma, Tombeck'bee, and Mississip'pi, whose course is *southerly.*

1. *All these rivers*, excepting the Tombeck'bee, and Mississippi, *rise* in the Al'leghany mountains.

2. The *Potomac, 3d class, flows* into the Chesapeak bay. It is *navigable* for large ships to Washington, 295 miles by the river and bay from the Atlantic, and for boats to Cumberland, nearly 200 miles above Washington.

3. *James River, 3d class*, is *navigable* for vessels of 120 tons to the falls, at Richmond, and for batteaux, 220 miles further.

4. The *Savannah, 4th class*, is *navigable* for large vessels to Savannah, 17 miles, and for boats, to Augusta.

Q. Which are the *chief towns ?*

A. *B...e* is the chief town of Maryland—*R...d* of Virginia—*W...n* of the District of Columbia—*N...n* of North Carolina—*C...n* of South Carolina—*S...h* of Georgia—*M...e* of Alabama—*N...z* of Mississippi and *N...w O...s* of Louisiana.

Q. Which are the *capitals* of these states ?

A. *A...s* is the capital of Maryland—*R...d* of Virginia—*W...n* of the District of Columbia—*R...h* of North Carolina—*C...a* of South Carolina—*M...e* of Georgia—*T...a* of Alabama—*J....n* of Mississippi—and *N...w O...s* of Louisiana.

Q. What *colleges* do the southern states contain ?

A. The *University of Virginia, Williams and Mary, Washington* and *Hampden Sidney* colleges in Virginia—*South Carolina* college in South Carolina—and *Franklin* college in Georgia.

Q. Which are the principal *religious denominations ?*

A. Methodists, Baptists, Presbyterians, Episcopalians, and Catholics.

Q. Which are the most *commercial towns* in the southern states ?

A. Balt'imore, Nor'folk, Charles'ton, Savan'nah, and New Or'leans.

Q. Which are the principal *productions* and *exports ?*

A. In the *northern* part of this division, *wheat, tobacco, and Indian corn;* in the *southern* part, cot'on, rice, and *sugar;* pitch, tar, and turpentine, are also exported in great quantities.

Q. What is the *character* of the people ?

A. In the higher classes, they are liberal, hospitable, independent in their feelings, and irascible; the lower classes are ignorant and vicious.

The greater part of this tract of country is *inhabited by* planters, having large plantations, and many slaves. They *live* at a considerable distance from each other—are *fond of* company and amusement, and especially of hunting, to which they frequently devote much time.

Questions.—*Map of the United States.*

Which are the Southern States ? Which is the largest of these states ? Which the second ? Third ? Which the least ? Where do the Alleghany mountains terminate ? Through what states do they run ? Between what states is Chesapeak Bay ? Between what capes does it open ? Where are Pamlico and Albemarle sounds ? What capes are near these sounds ? Is Cape Hatteras or Cape Lookout most northerly ? *What is the general course and place of discharge of the Potomac ? James ? Roanoke ? Pedee ? Santee ?*

Savannah? Alatamaha? Appalachicola? Alabama? and Tom-beckbee rivers? Point out the chief towns of the Southern States, *and tell where situated;* also point out the capitals, *and tell how situated?*

MARYLAND.

Q. How is Maryland *bounded?*

A. N. by p...; E. by d... and the ▲...; S. and W. by v...

1. *Maryland*, until 1632, *was considered a part of* Virginia; it began to be *settled* near the mouth of the Potomac, on the north side, in 1634, *by* Roman Catholics, under the direction of Lord Baltimore, the proprietor.

2. On the *eastern shore* the land is level; in the *middle*, hilly, and mountainous in the *west*. A sandy *soil* predominates in the *eastern* part, interspersed with rich meadows; in the *western* part, the vallies are fertile.

3. In the *eastern* section, the *climate* is moist and sickly; but in the *western*, is dry and salubrious.

4. *Length*, 196 miles; *breadth*, 120; *square miles*, 14,000. *Pop.* 407,350.

Q. Which is the principal *bay?*

A. Chesapeak bay, which lies chiefly within the boundaries of this state.

Q. Which are the principal *rivers?*

A. The Poto'mac and Susquehan'nah.

1. The *Potomac*, 3d class, *rises* in the Alleghany mountains, and *flows* southeasterly into Chesapeak bay. The *Susquehannah*, 4th class, *comes* from Pennsylvania, *flows* southerly, and also *enters* Chesapeak bay.

2. Those which *enter on the eastern shore*, are the Elk, Ches'ter, Chop'tank, Manticoke', Wicom'ico, and Pocomoke', all of which are of the 6th class, *rise* in Delaware, and have a southerly *direction*.

Q. Which is the *chief town*, and which the *capital?*

A. B...e is the chief town, and A...s the capital.

1. B...e, 4th class, *lies* on the Patapsco. It is the third city in the Union, in population and shipping.

2. *A...s, small town, is on* the Severn, 30 miles south of Baltimore.

3. The *other principal towns* are Fred'ericktown', Cumb'erland, and Hag'erstown, *small towns*.

Q. What *colleges* does this state contain?

A. The University of Maryland, St. Mary's, and Baltimore colleges, all in the city of Baltimore.

The *university* is yet in its infancy. The *other colleges* are flourishing, and respectable institutions.

Q. Which is the principal *religious denomination?*

A. The Roman Catholic.

Q. Which are its principal *manufactures?*

A. Flour is the principal; besides which are iron ware, glass, and paper.

Q. Which are its principal *productions* and *exports?*

A. Wheat, flour, and tobacco.

Questions.—Map of the United States.

How is Maryland bounded N. E. S. W.? What bay divides Maryland? What river comes from Pennsylvania and empties into Chesapeak bay? What river lies on the west? What cape at the southern extremity? What is the chief town, *and how situated?* What the capital, *and how situated?*

VIRGINIA.

Q. How is Virginia *bounded?*

A. N. by o..., p..., and m...; E. by m..., and the a...; S. by n... c..., and t...; W. by k..., and o...

1. The *first permanent English settlement* in N. America, *was made* in Virginia, in 1607, and was *called* Jamestown.

2. The *face of the country* along the *sea coast* is low, level, and swampy; in the *middle,* mountainous; in the *western* part, hilly.

3. The *soil* in the *eastern* part is sandy and barren; in the *interior,* alternately barren and fertile; *between* the *Blue Ridge* and the *Alleghany Mountains,* it is a fertile valley; *west* of this, it is generally barren.

4. The *climate,* on the *sea coast,* is hot and unhealthful in summers; but in the *upper country,* it is cool and salubrious.

5. *Length,* 370 miles; *greatest breadth,* 200; *square miles,* 64,000. *Pop.* 1,065,366.

Q. Which are the principal *mountains?*

A. The two ridges of the Alleghany mountains, known by the name of the Blue Ridge and North Mountains.

The *peaks of the Otter* in the Blue Ridge belong to the *6th class.*

Q. Which are the principal *rivers?*

A. The Poto'mac, James, Rappahan'nock, York, Ohi'o, and Great and Little Kenhawa, (Kennaw'wa.)

The *Potomac* and *James* rivers, *3d class*, the *latter* of which, *rises* west of the Blue Ridge, *flow* southeasterly into Chesapeak bay. The *Rappahannock* and *York*, 6th class, rise east of the Blue Ridge, and *flow* southeasterly into Chesapeak Bay. The *Ohio*, *2d class*, *runs* southwesterly, separating Virginia from the state of Ohio. The Great and Little Kenhawa, the former of which is of the *4th class*, the latter *6th class*, are *branches* of the Ohio, and have a *direction* north and northwest.

Q. What remarkable *swamp* does Virginia contain ?

A. Dismal Swamp.

This swamp is 30 miles *long*, and 10 *broad*. It *lies* partly in North Carolina.

Q. Which is the *chief town* and *capital* ?

A. R...d.

1. *R...d*, 6th class, *lies* on James River, 150 miles from its mouth. It is well situated for commerce.

2. The *other principal towns* are Nor'folk and Pe'tersburg, *large towns*, Fred'ericksburg, Wil'liamsburg, York'town, and Mount Ver'non, *small towns*.

3. *Yorktown*, on York river, is *famous* as the place where lord Cornwallis and his army were captured, on the 19th of October, 1781, by the united forces of France and America.

4. *Mount Vernon*, on the Potomac, *was formerly the seat of* General Washington, the friend, and father, and deliverer of his country.

Q. What *colleges* does Virginia contain ?

A. The University of Virginia, at Charlottesville; William and Mary college, at Williamsburg; Washington, at Lexington; and Hampden Sidney, in Prince Edward county.

This state has a *literary fund* of more than a million of dollars, the interest of which has been appropriated to common schools, and to the University of Virginia.

Q. Which are the principal *religious denominations* ?

A. Episcopalians were formerly the prevailing denomination; but at present Baptists, Methodists, and Presbyterians, are most numerous.

Q. What remarkable *mineral springs* exist here ?

A. The Hot springs in Bath county, and Berkley springs near the Potomac.

Q. What *minerals* are found in Virginia ?

A. Coal, near Richmond ; and iron, lead, and salt, west of the Blue Ridge.

Q. Which are the principal *productions* and *exports* ?

A. Wheat, flour, and tobacco.

Questions.—Map of the United States.

How is Virginia bounded N. E. S. W. ? What mountains pass through this state ? *Where do the Rappahannock, York, James, and Kenhawa rivers rise ? What is their course ? Where do they empty ?* Which is the chief town and capital ? *How situated?* What cape is at the S. E. corner ? *What town is situated near it ?*

DISTRICT OF COLUMBIA.

Q. How is the District of Columbia *situated ?*

A. It lies on both sides of the Potomac, 120 miles from its mouth.

1. This territory was *ceded,* in 1790, to the United States, and in 1800 *became the seat of the General Government.*

2. The *appearance* of the country is beautiful, and abounds in fine prospects.

3. The *soil* is light and sandy ; the *climate* is warm in summer, but occasionally cold in winter.

4. *Length,* 10 miles ; *breadth,* 10 ; *square miles,* 100. *Pop.* 33,039.

Q. Under whose *government* is it ?

A. Under that of the Congress of the United States.

Q. Which is the principal *river ?*

A. The Potomac.

Q. Which is the *chief town* and *capital ?*

A. W...n.

1. *W...n, 6th class,* is the capital of the United States. (*See page* 26.)

2. The *other towns* are Alexandria and Georgetown, *large towns.*

Q. What *colleges* does the District contain ?

A. A Roman Catholic college, at George'town, and the Columbian college, at Wash'ington.

NORTH CAROLINA.

Q. How is North Caroli'na *bounded?*

A. N. by v... ; E. by the A... ; S. by the A..., s... c..., and G... ; W. by s... c... and T...

1. The *territory*, now divided into the two Carolinas, Georgia, and Flor'ida, was *granted by* Charles II. king of England, *to* lord Clar'endon, and others, under whose direction the country was *settled*. A settlement, however, had been begun as early as 1650, in Albemarle county, by planters from Virginia *The province was divided into North and South Carolina, about* 1729.

2. *The face of the country* along the *sea coast* is low and level; as you proceed into the *interior*, the country becomes hilly and then mountainous.

3. The *soil* corresponds to the face of the country—poor along the sea coast, but improves as you proceed west.

4. The *climate* in the *eastern* part is hot and unhealthful, but is cooler and salubrious in the *west*.

5. *Length,* 430 miles ; *breadth,* 180 ; *square miles,* 48,000. *Pop.* 638,829.

Q. What *mountains* does this state contain ?

A. The Alleghany mountains cross the western part of the state.

Q. Which are the principal *rivers* ?

A. The Roanoke, Neuse, Pamlico, Cape Fear, Yad'kin, and Cataw'ba.

The *Roanoke,* 4*th class, rises* in Virginia, and *flowing* southeast, *enters* Albemarle sound ; *Neuse* river, 4*th class, rises* in this state, and *flowing* southeast, *enters* Pamlico sound ; *Pamlico* river, 6*th class, rises* in the northern part of the state, and *flowing* southeast, *empties* itself into Pamlico sound. *Cape Fear* river, 4*th class,* is *formed* of the Haw and Deep rivers, both of which *rise* in this state ; after their union the river has a *course* east of south, and *falls* into the ocean. The *Yadkin* and *Catawba rise* in the west, and *pass* into South Carolina, where the former is called the *Pedee'*, and the latter the *Wateree'*.

Q. , Which are the principal *sounds* ?

A. Al'bemarle and Pam'lico, which communicate with each other.

Q. Which are the principal *capes* ?

A. Cape Hat'teras, Lookout, and Fear—all dangerous to mariners, particularly C. Hatteras.

Q. Which is the *chief town*, and which the *capital* ?

A. N...n is the chief town ; R...h the capital.

1. *N...n, small town, lies* on the Neuse, and is a place of considerable commerce.

2. *R...h, small town,* is near the centre of the state.

3. The *other principal towns,* are Fay'etteville, Wil'mington, E'denton, and Plymouth, *small towns.*

Q. What *college* does the state contain ?

A. The University of North Carolina, at Chapel Hill, 28 miles from Raleigh.

Q. Which are the principal *religious denominations?*

A. Methodists and Baptists.

Q. Which are the principal *productions?*

A. Wheat, rye, barley, Indian corn, rice, and tobacco.

Q. Which are the principal *exports?*

A. Rice, tobacco, pitch, tar, turpentine, and lumber.

Questions.—Map of the United States.

How is N. Carolina bounded N. E. S. W.? *Describe the principal rivers?* What mountains pass through the western part of the state? What sounds can you mention ? What inlets? What capes? Which is the chief town? *How situated?* Which the capital? *How situated?*

SOUTH CAROLINA.

Q. How is South Caroli'na *bounded?*

A. N. by n... c...; E. by n... c... and the a...; S. by the a... and g...; W. by g...

1. The *face of the country* along the *sea coast,* or Lower country, as it is called, is low and swampy; the *Upper,* or *Back country,* is hilly and mountainous.

2. Much of the *soil* is sandy and barren, but with rich intervals.

3. The *climate* in the *low country* is hot and sickly; but in the *upper country* it is cool and healthful.

4. *Length,* 200 miles; *breadth,* 125; *square miles,* 28,000. *Pop.* 502,741.

Q Which are the principal *mountains?*

A. The Alleghany ridge crosses the northwestern part of the state.

Table mountain, the *highest peak,* belongs to the *6th class.*

Q. Which are the principal *islands?*

A. Sul'livan's, James, John's, and E'disto islands.

Q. Which is the principal *river?*

A. The Santee', is the great river of South Carolina.

1. The *Santee, 4th class,* is *formed* by the Congaree' and Wateree'. It *runs* in a southeast direction, and *falls* into the ocean.

2. The *other rivers,* both of which *run* southeasterly, and *fall* into the Atlantic, are the Pedee, *4th class,* which *rises* in North Carolina, where it is called the Yadkin; and the Edisto, *6th class.* The Saluda, *6th class,* is a *branch* of the Congaree.

Q. Which is the *chief town,* and which the *capital?*

A. C...n is the chief town, and C...a the capital.

1. *C...a, 5th class,* is *situated* between two small rivers, Ashley and Cooper, and is a place of much commerce. It has a sickly *climate.*

2. *C...a, small town, is on* the Congaree, 120 miles N. N. W. of Charleston, and has had a rapid growth.

3. The *other principal towns* are George'town, York, Cam'den, and Beaufort, (Bu'fort) *small towns.*

Q. What *college* does the state contain?

A. South Carolina college, at Columbia.

The state has appropriated 30,000 dollars, annually, for the *support of free schools.*

Q. Which are the principal *religious denominations?*

A. Methodists and Baptists.

Q. Which are the principal *productions?*

A. Cotton and rice.

Q. Which are the principal *exports?*

A. Cotton and rice; besides which are lumber, pitch, tar, &c.

Questions.—Map of the United States.

How is S. Carolina bounded N. E. S. W.? In what part of the state is Table mountain? *Describe the rivers.* What islands lie off the coast? Which is the chief town? *How situated?* Which the capital? *Which way from Charleston?*

GEORGIA.

Q. How is Georgia *bounded?*

A. N. by t... and n... c...; E. by s... c... and the a...; S. by f...; W. by the a...

1. The *settlement* of Georgia was *begun* in 1733, by 116 planters from England, *under* Gen. Og'lethorpe. *They built* Savannah.

2. The *face of the country* and *climate* strongly resemble those of the Carolinas.

3. The *soil* is generally fertile.

4. *Length,* 270 miles; *breadth,* 250; *square miles,* 60,000. *Pop.* 340,989.

Q. Which are the principal *rivers ?*

A. The Savan'nah and Alatamaha, (Al'ta-ma-haw'.)

1. *Savannah river, 4th class, is formed* by the union of two branches, both of which *rise* in the western part of North Carolina. The river *separates* Georgia from S. Carolina, and *runs* in a southeast direction into the Atlantic. The *Alatamaha,* 3d *class,* is *formed* by the Oconee' and Oak-mulgee', which *rise* in the northern part of the state. After their union, the river *runs* southeast, and *empties* into the Atlantic.

2. The *other principal rivers,* all of which *flow* southerly, are the Ogechee, 5*th class,* and the Chatahoo'chee, and Flint rivers, which last *unite* at the S. W. extremity of the state, and form the Apalachicola.

Q. What remarkable *swamp* does the state contain ?

A. Okefono'co swamp.

This swamp *lies* partly in Florida. It is 180 miles in *cir-cumference,* and is *full of* alligators, snakes, &c.

Q. Which is the *chief town,* and which the *capital ?*

A. S...h is the chief town, and M...e the capi-tal.

1. *S...h, large town, is on* Savannah river, 17 miles from the mouth. It is a place of much business.

2. *M...e, small town, is on* the Oconee, near the centre of the state, 300 miles from the ocean.

3. The *other principal towns,* are Augus'ta, Da'rien, Sun'-bury, Pe'tersburg, and Ath'ens, *small towns.*

Q. What *college.* does the state contain ?

A. Franklin college, at Athens.

Q. What *Indians* live in this state ?

A. The Creek Indians.

The *Creeks occupy* the western part of this state, and the eastern part of Alaba'ma. Their *number* is about 20,000. They are a warlike and powerful tribe. The *Cher'okees inha-bit* the northwestern corner of Georgia, and the adjacent parts of Alaba'ma and Tennessee', and are *estimated* at about 12,000.

Q. Which are the principal *religious denominations ?*

A. Baptists and Methodists.

Q. Which are the principal *productions* and *exports?*

A. Cotton, rice, and tobacco.

Questions.—*Map of the United States.*

How is Georgia bounded N. E. S. W.? *Where does the Savannah river rise? What is its course? Where does it empty? Describe in like manner the Alatamaha and Chatahoochee. What part of the state do the Cherokees and Creek Indians inhabit?* Which is the chief town? *How situated?* Which the capital? *How situated?*

ALABAMA.

Q. How is Alabama *bounded?*

A. N. by T...; E. by G...; S. by F..., and the G. of M...; W. by M...

1. Alabama has been *settled* since the American revolution. In 1800, it *was included* in the territorial government of Mississippi. In 1817, it was *separated* from Mississippi, and became a territory by itself. In 1819, it was *admitted into the union* as a state.

2. The *face of the country* in the *south*, is level, but rises into hills in the *interior*, and in the *north* into mountains.— The *soil* is good.

3. A mild *climate* prevails in winter, and the *heat of summer is moderated by* the refreshing breezes of the Gulf of Mexico.

4. *Length*, 317 miles; *breadth*, 174; *square miles*, 46,000. *Pop.* 144,317.

Q. Which are the principal *rivers?*

A. Chatahoo'chee, Tennessee', Appalachico'la, and Mobile, (Mo-beel') with its branches.

1. The *Chatahoochee rises* in the northern part of Georgia, and *runs* first in a southwest direction, till it reaches the eastern boundary of Alabama, where it *turns* east of south, and, at length, *uniting* with Flint river, *forms* the Appalachicola. The Tennessee, *2d class*, only *enters* Alabama, curves, and leaves it.

2. *Mobile empties* into Mobile bay. It is *formed* by the union of the Alabama and Tombeckbee, the former of which *runs* southwesterly, and the latter southerly. The *Alabama*, *4th class*, is *formed* by the Coo'sa and Tallapoo'sa, both of which *rise* in Georgia. The *Tombeckbee, 4th class, rises* in the northwestern part of the state. The *Blackwarrior, 5th class, rises* in this state, and pursuing a southwest *course, empties* into the Tombeck'bee.

Q. Which is the *chief town*, and which the *capital?*

A. M...e is the chief town; T...a the capital.

1. *M...e, small town, is on* the west side of Mobile bay. It is a place of considerable trade.

2. *T...a, small town,* is *situated* on the Tuscaloosa river, north of Mobile.

3 The other *principal towns,* are Blakely, St. Stephens, Huntsville, and Cahawba, *small towns.*

Q. What *Indians* live in this state?

A. The Creeks, Cherokees, and Choctaws.

The *Creeks* occupy the *eastern;* the *Cherokees* the *northern;* and the *Choctaws* the *western* part of the state.

Q. What *forts* are situated in the state?

A. Fort Stod'dard on Mobile river; fort Clai'borne on the Alabama; and fort Jackson, near the junction of the Coo'sa and Tallapoo'sa.

Q. Which is the principal *production?*

A. Cotton. Rice, corn, and wheat are, however cultivated.

Q. What is the principal article of *export?*

A. Cotton.

Questions.—*Map of the United States.*

How is Alabama bounded N. E. S. W.? What mountains terminate in this state? *Describe the Chatahoochee. Tennessee, Appalachicola, and Mobile rivers.* Which is the chief town? Which the capital? *How are they situated?*

MISSISSIPPI.

Q. How is Mississippi *bounded?*

A. N. by т...; E. by а...; S. by the ɢ... of м..., and ʟ...; W. by ʟ..., and а... т...

1. As early as 1716, the French formed a *settlement,* where Natchez now stands, and built a fort. In 1763, the *French ceded the country to the English*—in 1793, the *English ceded it to Spain;* in 1798, *Spain abandoned it to the United States.* In 1800, Mississippi *became a territory,* and in 1817, *was erected into a state.*

2. The *face of the country* in the *southern* part of the state is level; but is elevated towards the *north.* The *soil* is generally good, and in many places very fertile.

3. The *climate* is mild in winter, and less hot in summer, than in many places in the same latitude.

4. *Length,* 339 miles; *breadth,* 150; *square miles,* 45,700. *Pop.* 75,448.

Q. Which are the principal *rivers?*

A. The Mississippi, Yazoo', Black, Pearl, and Pascagou'la.

The *Mississippi*, 1*st class*, is the western boundary. The *Yazoo*, 5*th class*, and *Black* rivers, 6*th class*, both of which *rise* in this state, and *run* southwesterly, *are branches* of the Mississippi. *Pearl* river, 5*th class*, *rises* in the Choctaw country, and *running* southerly, *empties* into a lake which communicates with the Gulf of Mexico. *Pascagoula*, 5*th class*, *rises* in the same country, and *flows* southerly into the Gulf of Mexico, 40 miles west of Mobile bay.

Q. Which is the *chief town*, and which the *capital?*

A. N...z is the chief town, and J....n the capital.

1. *N...z, small town, is on* the Mississippi, 320 miles by the river, and 156 by land, above New Orleans. It is the *place of deposit* for the merchandise of the western part of the state.

2. *I...n, small town*, is in the central part of this state, northeast of Natchez.

3. The other *principal towns* are Washington, Palmyra, and Monticello, *small towns.*

Q. What *Indians* inhabit this state?

A. The Choctaw and Chickasaw tribes.

1. The *Choctaw* tribe *inhabits* the central part of the state. They are estimated at 20,000. A *missionary station* is located at Elliot, in their country, which is *under the patronage of* the American Board of Commissioners for Foreign Missions. The Indians treat the missionaries kindly.

2. The *Chickasaws live* in the *northern* part of this state, and the adjacent parts of Tennessee. They are partly civilized. Their *number* is about 6500.

Q. Which are the principal *productions* and *exports?*

A. Cotton and rice.

Questions.—*Map of the United States.*

How is Mississippi bounded N. E. S. W.? *Describe the rivers Mississippi, Yazoo, Black, Pearl, and Pascagoula. Where do you find the Chickasaws and Choctaws?* Which is the chief town, and which the capital? *How are they situated?*

LOUISIANA.

Q. How is Louisian'a *bounded?*

A. N. by A... T..., and M...; E. by M..., and the G... of M...; S. by the G... of M...; W. by M...

1. *Louisiana* formerly *embraced* the whole *territory* between the Mississippi river and the Pacific ocean, and was *owned* by France. In 1803, that *power sold* it to the United

States, *for* $15,000,000, since which time, it has been *divided into* Missouri Territory, state of Missouri, Arkansas Territory, and state of Louisiana. Louisiana *began to be settled* as early as 1699, *by the French.* It was *admitted into the union* in 1811.

2. *The country on the Gulf of Mexico consists* of low meadow land; about the *mouth of the Mississippi,* for many miles, it is a continued swamp. One fifth of the state is sometimes *inundated.*—The *soil* is fertile.

3. The *climate* of the southern part is hot, and sickly in the summer; the winters are mild.

4. *Length,* 240 miles; *breadth,* 210; *square miles,* 48,220. *Pop.* 153,407.

Q. Which are the principal *rivers?*

A. The Mississippi, Red, and Wash'ita.

1. The *Mississippi, 1st class,* is the eastern boundary of the state, and *discharges* itself 100 miles below New Orleans. *Red river, 2nd class, rises* in the Spanish dominions among the Rocky mountains, and after *running* 1200 miles east and south, *enters* the Mississippi. The *Washita, 3d class, rises* in Arkansas Territory, and *flows* south into Red river.

2. Besides these there are several *other considerable rivers,* the Pearl, Sabine, Atch'afalay'a, Ib'erville, &c.

Q. Which is the principal *lake?*

A. Pontchartrain'.

Q. What do you understand by the *Levees?*

A. They are banks erected along the sides of rivers, to prevent the water from overflowing the plantations.

A levee, 100 miles long, has been erected along the Mississippi, *above, and below New Orleans.* These banks, though strong, are sometimes broken by the water, when whole plantations are ruined.

Q. Which is the *chief town* and *capital?*

A. N...w O...s.

1. *N...w O...s, 5th class, stands on* an island formed by the Mississippi and Iberville, 100 *miles above the mouth* of the former. It is well *situated for trade,* and is already a great city. *It is liable,* however, *to* great mortality during the sickly months.

2. The *other principal towns* are Natchitoches, (Nak'e-tosh) Opelousas, (Op-pe-loo'suz) Alexandria, Baton Rouge, (Bat'-ton Ruzhe') *small towns.*

Q. Which is the principal *religious denomination?*

A. The Roman Catholic.

Q. Which are the principal *productions* and *exports ?*
A. Cotton, sugar, and rice.

Questions.—Map of the United States.

How is Louisiana bounded N. E. S. W. ? *Describe the Mississip pi, Red, and Washita rivers.* What lake do you notice ? What river separates Louisiana from the Texas ? Which is the chief town and capital, *and how situated ? Where is Natchitoches ? Where Alexandria ? What is the latitude and longitude of New Orleans ? Which way from N. Orleans is New York ?*

WESTERN STATES.

Q. Which are the *western states ?*
A. Tennessee', Kentucky, Ohi'o, Indiana, (In-je-an'na) Illinois, (Il-le-noy') and Missouri, (Mis-soo're.)

1. The *eastern parts* of Tennessee and Kentucky are mountainous; the *rest* of the country is, for the most part, but moderately uneven. *Between the Mississippi* and *Rocky mountains are to be found* immense prairies, or tracts of land entirely destitute of trees, and covered with grass, which, in some places, is more than 6 feet high.
2. The *soil* of the western states is generally very fertile.
3. The *climate* is milder than that of the states, in the same latitude, on the Atlantic ocean.

Q. Which are the principal *mountains ?*
A. The Cumberland mountains, which divide Tennessee into Eastern and Western Tennessee, and the Alleghany ridge, which separates Tennessee from North Carolina.

Q. Which are the principal *rivers?*
A. The Mississippi, with its tributary streams, the Ohio, Tennessee, Cumberland, and Illinois.

1. The *Ohio, 2d class*, is *formed by* the Monongahela and Alleghany, which *unite* at Pittsburg. It *flows* in a south-westerly direction, and *enters* the Mississippi a little above the southern boundary of Kentucky. At some seasons, *vessels* of 200 or 300 tons *descend the river from Pittsburg.*
2. The *Tennessee, 2nd class*, whose *course* is first south-westerly, and then northwesterly, is *navigable* for boats, throughout nearly its whole extent.
3. The *Cumberland, 3d class,* also has a southwesterly,

and afterwards a northwesterly *direction*, and is *navigable* for large vessels to Nashville, 200 miles, and for large boats 500 miles.

4. The *Illinois, 4th class*, whose *course* is very circuitous, though in general southerly, is a fine river, *flowing* into the Mississippi, 21 miles above the junction of the Missouri. It is *navigable* for boats almost to lake Michigan.

Q. Which are the *chief towns* ?

A. *N...e* is the chief town of Tennessee—*L...n* of Kentucky—*C...i* of Ohio—*V...s* of Indiana—*K...a* of Illinois—and *St. L...s* of Missouri.

Q. Which are the *capitals* of these states ?

A. *M...h* is the capital of Tennessee—*F...t* of Kentucky—*C...s* of Ohio—*I...s* of Indiana—*V...a* of Illinois—and *J...n* of Missouri.

Q. What *colleges* does this state contain ?

A. A college at *Green'ville* and *Knox'ville*, in Tennessee—*Transylva'nia* university in Kentucky—and a college at *Ath'ens* and *Cincinna'ti*, Ohio.

Q. Which are the principal *religious denominations?*

A. Presbyterians, Methodists, and Baptists.

Q. What *antiquities* deserve notice ?

A. Mounds of earth, fortifications, and walls of various forms and dimensions, the work of some ancient people, now unknown.

These monuments are found scattered from the great lakes to the gulf of Mexico, and from the Alleghany mountains to the Pacific ocean. They *indicate* great labour. Trees several hundred years old, *are often seen growing out of them.* Their *origin* and *history* are now unknown.

Q. What can you say of the *commerce* of the western states ?

A. It centres chiefly at New Orleans, although considerable trade is carried on with New York, Philadelphia, and Baltimore.

Q. Which are the principal *productions ?*

A. Wheat, Indian corn, rye, cotton, and tobacco.

Q. What is the *character* of the people ?

A. They are hardy, brave, and industrious; but the lower classes are uneducated and unrefined.

Questions.—Map of the United States.

Which are the Western States? Which is the largest? Which the smallest? Which states lie west of the Mississippi? Which of the western states lie east? What states does the Ohio river separate? What states does the Mississippi separate? Through what states do the Cumberland mountains pass? *What is the direction and place of discharge of the Ohio? Kentucky? Green? Scioto? Miami? Wabash? Tennessee? Cumberland? and Illinois?* Which are the chief towns, *and how situated?* Which are the capitals, *and how situated?*

TENNESSEE.

Q. How is Tennessee' *bounded?*

A. N. by K..., and v...; E. by N... c.,..; S. by G..., A..., and M... ; and W. by A... T..., and M...

1. *Tennessee began to be settled* in 1765; and in 1773 the inhabitants had considerably increased. Until 1790, it was a *part of North Carolina;* but at that time *it was ceded to* the United States, and with some other tracts, had a territorial government. It 1796, it *became* a *state.*

2. The *eastern* part is mountainous; the *western,* partly level and partly hilly. *East Tennessee* has a barren *soil;* the *soil of West Tennessee* is fertile.

3. The *climate* of the eastern part is temperate; the *summers of the western part* are hot; but the *winters* are mild.

4. *Length,* 420 miles; *breadth,* 102; *square miles,* 40,000. *Pop.* 422,813.

Q. Which are the principal *mountains?*

A. The Cumberland mountains, which divide the state in the middle, into East and West Tennessee'.

Q. Which are the principal *rivers?*

A. The Mississip'pi, Cum'berland, and Tennessee'.

1. The *Mississippi,* 1st class, is the western boundary. The *Cumberland,* 3d class, *rises* in Kentucky, *runs* southwesterly into Tennessee, and again passes by a northwest *course* into Kentucky, where it *flows* into the Ohio. The *Tennessee,* 2nd class, *rises* in Virginia *under the name of* Holston, *runs* into Tennessee, thence southwesterly into Alabama, and again northwesterly into Tennessee, whence it passes into Kentucky, where it *unites* with the Ohio.

2. The *branches* of the Tennessee are the Hiwassee', Clinch, and Duck.

Q. Which is the *chief town,* and which the *capital?*

A. N...e is the chief town; M...h is the capital.

1. *N...e, small town, is on* the Cumberland. It has a flourishing trade.

2. *M...h, small town, is near* the centre of the state, 32 miles S. E. of Nashville.

3. The *other principal towns*, are Knoxville, Clarksville, Washington, *small towns.*

Q. What *Indians* live partly in this state ?

A. The Cherokees'.

The *number* of *Cherokees* is about 12,000. They are partially civilized. A *missionary station* has been located at Brainerd, and schools in several other places.

Q. What *colleges* does Tennessee contain ?

A. One at Greenville, in East Tennessee, and another at Nashville, in West Tennessee.

Q. Which are the principal *religious denominations* ?

A. Methodists, Baptists, and Presbyterians.

Q. Which are the principal *productions* and *exports* ?

A. Cotton, tobacco, and hemp.

Questions.—Map of the United States.

How is Tennessee bounded N. E. S. W. ? What mountains pass through this state ? *Where does Tennessee river rise ? Describe its windings. What mountains separate it from Cumberland river ? Where does it empty ? Describe in like manner the Cumberland river.* Which is the chief town ? Which the capital ? *How are they situated ?*

KENTUCKY.

Q. How is Kentucky *bounded* ?

A. N. by i..., i..., and o...; E. by v...; S. by t...; and W. by m..., i..., and i...

1. *Kentucky began to be settled* in 1775, by a number of families *under the direction* of Col. Daniel Boone. The first settlers suffered much from the Indians. The *territory belonged to* Virginia, and continued under her jurisdiction till 1790, when it *became an independent state.*

2. The state is *mountainous* in the eastern part. On the Ohio river it is *hilly*, but the rest of the state is more *level.*

3. The *soil* is poor in the *eastern* part, but fertile in the *interior* and *western* part.

4. The *climate* is temperate and healthful.

5. *Length*, 300 miles; *breadth*, from 40 to 180; *square miles*, 42,000. *Pop.* 564,317.

Q. Which are the principal *rivers* ?

A. Kentucky, Licking, Green, Cumberland, and Tennessee.

The *Kentucky*, *5th class*, *rises* in the Cumberland mountains, and *running* in a northwest direction, *empties* itself into the Ohio, 77 miles above the rapids, at Louisville. *Licking* river, *6th class*, *rises* in the Cumberland mountains, and *flowing* northwest, *empties* itself into the Ohio, opposite Cincinnati. *Green* river, *5th class*, *rises* in the centre of the state, and *flowing* westerly, *empties* itself into the Ohio, 50 miles above the mouth of the Cumberland.

Q. Which is the *chief town*, and which the *capital?*

A. L...n is the chief town ; F...t the capital.

1. *L...n, small town*, is *situated* 30 miles S. E. of Frankfort The town was laid out in 1782.

2. *F...t, small town, lies* on Kentucky river.

3. The *other principal towns* are Louisville, (Loo'eville) New'port, and Mays'ville, *small towns.*

Q. What *college* does this state contain ?

A. Transylvania university, at Lexington.

Q. Which are the principal *religious denominations* ?

A. Baptists, Presbyterians, and Methodists.

Q. What can you say of the *salt springs* ?

A. Salt springs abound, from which sufficient salt is made to supply Kentucky, and a great part of Ohio and Tennessee.

Q. Which are the principal *productions* and *exports* ?

A. Tobacco, hemp, and wheat.

Questions.—*Map of the United States.*

How is Kentucky bounded N. E. S. W. ? *Describe the rivers, where they rise ; what is their course ; and where they empty.* Which is the chief town, and which the capital ? *How are they situated ?*

OHIO.

Q. How is Ohio *bounded ?*

A. N. by M... T..., and L... E... which separates it from U... C... ; E. by P..., and V... ; S. by V..., and K... ; W. by I...

1. Until 1787, Ohio *was inhabited* principally by Indians, a few Moravians, and trespassers on public lands. In 1788, the *first settlement was begun* at Mariet'ta, *under* Gen. Rufus Putnam, from New England. In 1805, it was *admitted into the union*, by congress, *as an independent state.*

2. The *southern part* of the state is hilly ; the *rest* is more level, and the *soil* is in general fertile.

3. The *climate* is temperate, and more healthful than in former years.

4. *Length,* 216 miles ; *breadth,* 216 ; *square miles,* 39,000
Pop. 581,434.

Q. Which is the principal *river ?*

A. The Ohio.

1. The *Ohio, 2nd class,* is *formed* by the Alleghany and Monongahela, the *former* of which *rises* in Pennsylvania ; the *latter* in Virginia.   They *unite* at Pittsburg, whence *proceeding* southwesterly, the river *runs* along the whole southern border of the state, a distance of 480 miles, *separating it from* Virginia and Kentucky, and at length *falls* into the Mississippi.

2. The *principal rivers* which *fall into the Ohio,* are the Muskin'gum, 5*th class,* Hockhock'ing, Scio'to, and Great Miam'i, all of the 6*th class,* and all of which have a southerly *direction.*

3. Those which *fall into Lake Erie* by a northerly *course,* are the Maumee', Sandus'ky, and the Cuyaho'ga, all of the 6*th class.*

Q. What *lake* borders upon this state ?

A. Lake Erie on the North.

Q Which is the *chief town,* and which the *capital ?*

A. C...i is the chief town ; C...s the capital.

1. C...i, *large town,* lies on the Ohio river, in the southwestern corner of the state.   It *contains* extensive manufacturing establishments.

2. C...s, *small town, is on* the Scioto, near the centre of the state.

3. The *other principal towns* are Chilico'the, Zanes'ville Mariet'ta, Ath'ens, and Cleve'land, *small towns.*

Q. Which are the principal *religious denominations ?*

A. Presbyterians and Methodists.

Q. What *minerals* have been noticed ?

A. Coal and iron ore.   Salt springs have also been discovered.

Q. What *curiosities* does the state contain ?

A. Extensive ancient fortifications, at Marietta.

By what people these fortifications were constructed, or *at what period,* is wholly unknown.

Q. Which are the principal *productions* and *exports ?*

A. Wheat, flour, hemp, and flax.

Questions.—*Map of the United States.*

How is Ohio bounded N. E. S. W. ? *Describe the rivers which fall into the Ohio, and those which fall into Lake Erie.* Which is the chief town ? Which the capital ? *How are they situated ? Where is Cleveland ?*

INDIANA.

Q. How is Indiana (In-je-an'na) *bounded ?*

A. N. by L. M..., and M... T...; E. by o..., and
x...; S. by k...; W. by i...

1. The exact period *when Indiana began to be settled* is
uncertain, although some French families were found in the
territory upwards of a century ago.   Until 1801, it *formed a
part of the great northwestern territory ;* but at that date it
was *erected into a territorial* government. In 1816, it was
admitted as a state into the union.

2. The *face of the country* near the Ohio, is hilly ; further
north it is level, and *abounds* in extensive and fertile prairies.

3. The *climate* is temperate, and generally healthful.

4. *Length,* 284 miles ; *breadth,* 155 ; *square miles,* 37,000
Pop. 147,178.

Q. Which are the principal *rivers ?*

A. The Ohio and Wabash, (Wau'bosh.)

1. The *Ohio* is the southern boundary of the state from the
mouth of the great Miami to that of the Wabash.

2. The *Wabash,* 3d class, *rises* in the northeastern part of
the state, and *flows* southwesterly into the Ohio, 30 miles
above the mouth of the Cumberland.

3. Tippecanoe, (Tip-pe-ca-noo') and White River, both of
the *6th class,* are *branches* of the Wabash.   The *former rises*
in the northern part of the state, and *runs* in a southerly direction ; the *latter rises* in the eastern part of the state, and
runs in a southwesterly direction.

Q. Which is the *chief town,* and which the *capital ?*

A. V...s is the chief town ; I...s is the capital.

1. *V...s, small town, is on* the Wabash, 100 miles from its
mouth.   It was settled by the French, in 1730.

2. *I...s, small town, is* about 110 miles north of Louisville,
Kentucky.

3. The *other principal towns* are Vevay, Clarksville, and
Madison, *small towns.*

Q. Which are the principal *productions?*

A. Indian corn, wheat, and cotton.

The *vine is cultivated by* some Swiss settlers at Vevay, and
from 5 to 8000 *gallons of wine are annually made.*

Questions.—*Map of the United States.*

How is Indiana bounded N. E. S. W. ? What are its rivers ?
Where do they rise? What is their course ? Where do they empty?
Which is the chief town ? Which the capital ? *How situated?*

ILLINOIS.

Q. How is Illinois (Il-le-noy') *bounded ?*

A. N. by the N... W... T...; E. by M... T..., and I...; S. by K... and M...; W. by M... and M... T...

1. Illinois *derives its name from* its principal river, which, in the language of the Indians, *signifies* the river of men. The first *settlements* were those of Kaskaskia and Cahokia, by the French, some time before the 18th century. Illinois was a *part of Indiana, until* 1809, when it was erected into a territorial government. In 1818 it *became an independent state.*

2. The *face of the country* is generally flat; but hilly towards the *north.* It *abounds* in extensive prairies, and is very fertile.

3. The *climate* is agreeable, but unhealthful in some districts.

4. *Length,* 345 miles; *breadth,* 210; *square miles,* 52.000. *Pop.* in 1826, 72,817.

Q. Which are the principal *rivers ?*

A. Mississippi, Ohio, Wabash, Illinois', Kaskaskia, and Rock river.

1. The *Mississippi, Ohio,* and *Wabash,* are boundary rivers on the W., S., and E., for more than 1000 miles.

2. *Illinois, 4th class,* which is the principal river, *rises* in Indiana, and *flows* southwesterly into the Mississippi, 21 miles above the Missouri.

3. *Kaskaskia, 6th class, rises* in the eastern part of the state, and *flows* southwesterly into the Mississippi, 130 miles above the Ohio. *Rock* river, *5th class, rises* in the northern part of the state, and *enters* the Mississippi, by a southwest *course,* 160 miles above the Illinois.

Q. Which is the *chief town,* and which the *capital?*

A. K...a is the chief town; V...a the capital.

1. *K...a, small town, is on* Kaskaskia river, 11 miles from its mouth. It was settled by the French more than 100 years ago.

2. *V...a, small town, is on* the same river, 70 miles above St. Louis.

3. The *other principal towns* are Caho'kia, Shawnee', and Ea'wardsville, *small towns.*

Q. What *minerals* does the state contain ?

A. Coal, copper, lead, iron, and salt.

Extensive salt works are established 12 miles west of Shaw-

7

nestown.   Between 200,000 and 300,000 *bushels are annu*
made at these works.

Q.   What has been done for the *support of schools ?*

A. Congress has granted one section in every
township for the support of common schools, and two
townships for a university.

Q.   Which are the principal *productions?*

A. Corn, wheat, flax, hemp, and cotton.

Questions.—*Map of the United States.*

How is Illinois bounded N. E. S. W.?   Which are its rivers?
Where do they rise?   What is their course?   Where do they empty?
Which is the chief town?   Which the capital?   *How are they*
situated?

MISSOURI.

Q.   How is Missouri (Mis-soo'ree) *bounded?*

A. N. by M... T..., and I...; E. by I..., K..., and
T...; S. by A... T...; W. by M... T...

1. Missouri *began to be settled* soon after the peace of 1763
St. Louis was settled in 1764.   Missouri was for a time connected with Alabama in a territorial government; but in 1819,
became an independent state.

2. The *face of the country* is generally level, and the *soil*
fertile.

3. The *climate* is generally pleasant and healthful.

4. *Mean length,* 280 miles; *mean breadth,* 220; *square*
miles, 63,000.   *Pop.* 66,586.

Q.   Which is the principal *river?*

A. The Missouri.

1. The Missouri, *1st class, passes* through the centre of the
state, in an easterly *direction,* and *enters* the Mississippi 20
miles below the mouth of the Illinois.

2. The Gasconade, *5th class,* Osage, (Waw-sash'y) *3d class,*
and Grand river, *3d class,* are *branches* of the Missouri.

Q.   Which is the *chief town,* and which the *capital?*

St. L...s is the chief town; J...n the capital.

1. *St. L...s, small town, is on* the Mississippi, 14 miles by
land, below the mouth of the Missouri.   It is well situated for
commerce.

2. *J...n, small town,* is a new town, *on the* Missouri, a few
miles above the mouth of the Osage.

3. The *other principal towns* are St. Genevieve, St. Charles,
Cape Girardeau, and New Madrid, *small towns.*

Q. What *mineral* abounds in Missouri ?

A. Lead.

The *mines lie* about 40 miles west of the Mississippi.—The mine district *contains upwards* of 3000 *square miles.*—There are 45 *mines*, which employ 4100 *persons*, and *yield* annually 3 or 4,000,000 *pounds of lead.*

Q. Which are the principal *productions ?*

A. Indian corn, wheat, and cotton.

Q. Which the principal *exports ?*

A. Lead and furs.

Questions.—Map of the United States.

How is Missouri bounded N. E. S. W. ? What river passes through it ? What are the other rivers ? Which is the chief town, and which the capital ? *How situated?*

TERRITORIES.

Q. What *Territories* belong to the United States ?

A. Michigan, North Western, Missouri, Western, Arkansas, (Ar-kan-saw') and Florida.

MICHIGAN TERRITORY.

Q. How is Michigan Territory *bounded?*

A. N. by L... s..., Strait of St. M... and L... H..., which separate it from U... C...; E. by U... C... from which it is separated by L... H..., St. C... and E...; S. by O... and I... ; W. by I... and N... W... T...

1. This territory was *formerly inhabited* by the Hurons, a powerful tribe of Indians. It *belonged* to the Great North Western Territory until 1805, when it was detached and erected into a distinct territorial government. Gen. Hull was the *first governor.*

2. The *face of the country* is elevated in the middle, with a gradual descent in every direction. The *soil* is rich.

3. The *climate* is temperate in the south ; but colder in the north.

4. *Length*, 286 miles ; greatest *breadth*, 174 ; *square miles*, 34,600. *Pop.* 8896.

Q. Which are the principal *bays* in Michigan ?

A. Saganaw', on the west of Lake Huron, Green Bay, on the northwest, and Traverse Bay, on the northeast of lake Michigan.

Q. What *lakes* lie in this state ?

A. Lake Mich'igan lies wholly in this state ; about

half of lakes Huron and St. Clair. Lakes Superioi and Erie touch upon this territory.

Q. Which are the principal *rivers* ?

A. Michillimackinac, (Mack'e-naw) St. Clair, Detroit, and Saganaw, which enters Saganaw Bay.

Michillimackinac connects lake Michigan with lake Huron. *St. Clair connects* lake Huron and lake St. Clair.—*Detroit connects* lake St. Clair and lake Erie. *Saganaw enters* Saganaw bay, *all of the 6th class.*

Q. Which is the *chief town* and *capital ?*

A. D...t.

1. *D...t, small town, is on* Detroit river, *between* lake Erie and lake St. Clair.

2. The *other principal towns* are Brownstown and Michillimackinac, *small towns.*

Q. What tribes of *Indians* inhabit this territory ?

A. The Chippewas, Ottawas, Potowottamies, Wyandots, Munsees, Shawnese, and Delawares.

Q. What can you say of its *inland navigation ?*

A. Steam-boats go regularly, during the summer, from Detroit to Buffalo.

Q. Which are its principal *productions ?*

A. Wheat, rye, oats, and barley.

Questions.—Map of the United States.

How is Michigan bounded N. E. S. W. ? *Where does Saganaw river empty ?* Which side of lake Michigan is Green Bay ? Which side is Traverse Bay ? What lake lies wholly in this Territory ? Where is the strait of St. Mary ? Of Michillimackinac ? Which is the chief town and capital ?

NORTH WESTERN TERRITORY.

Q. How is the North Western Territory *bounded ?*

A. N. by u... c..., and l... s... ; E. by l... s... and m... t...; S. by i..., and m... t... ; W. by m... t...

1. This territory has no *government* of its own, but is *incorporated with the government* of Michigan, and *forms a county by the name of* Crawford.

2. The *face of the country* is sometimes hilly, but generally level, and the *soil* good.

3. The *climate* in winter is very severe, but temperate in summer, and at all seasons may be said to be healthful.

4. *Length*, 450 miles; *breadth*, 350; *square miles*, 140,000. *Pop.* about 1000.

Q. Which are the principal *rivers?*

A. The Mississippi, Ouisconsin, (Wees-con'sin) Chippeway, St. Croix, and Fox.

The Mississippi is *the western boundary:* all the other rivers *enter* the Mississippi, excepting Fox river, which *falls* into Green bay.

Q. What are the principal *productions?*

A. A kind of rice which grows wild.

Questions.—*Map of the United States.*

How is the North West Territory bounded N. E. S. W.? Where is Fox river? *Which way does it flow? Where does it empty? Describe in like manner the Ouisconsin, Chippeway, and St. Croix.*

MISSOURI TERRITORY.

Q. How is Missouri Territory *bounded?*

A. N. by the British Dominions; E. by the N... w... T... and the States of I... and M...; S. by the T... of A... and the Spanish Dominions; and W. by the w... T...

1. The *eastern part* of the country is generally level, and has extensive prairies of fine *soil. Towards the west,* the country becomes elevated and mountainous.

2. The *square miles* are estimated at 800,000.

Q. Which are the principal *mountains?*

A. The Rocky, or Stony Mountains.

The *Rocky* mountains, *5th class, run* from S. E. to N. W. They *rise* abruptly from the plains on the eastern side, and are always covered with snow. These mountains *divide the waters* which flow into the Pacific, from those which flow into the Atlantic.

Q. What is the principal *river?*

A. The Missouri.

1. The *Missouri, 1st class, rises* on the east side of the Rocky mountains. Its *course* is first northerly, then easterly, south, and southeasterly, until, at length, it *enters* the Mississippi, above St. Louis. Its *principal tributaries* are the Yellowstone, Platte, and Kansas, all of the *2d class,* and all *rise* in the Rocky mountains. The *course* of the Yellowstone is northeasterly; the *course* of the others is easterly.

Q. What *Indian tribes* inhabit these regions ?

A. The Sioux, Osages, Kansas, and Pawnees ; but little is known about them.

The Sioux are the most *powerful tribe* in N. America.—Their *number* has been *estimated* at 22,000. The *whole number* of Indians in this territory is *supposed to be* nearly 150,000.

Q. What *military posts* have the United States in this territory ?

A. One at Council Bluff, on the east side of the Missouri, above the mouth of the Platte ; and another at St. Peter's river.

Q. What *animals* are found here ?

A. Buffaloes, which are hunted for their hides and tallow ; also bears, deer, elk, and panthers.

Questions.—*Map of the United States.*

How is Missouri Territory bounded N. E. S. W. ? What range of mountains is found here ? In what direction do they run ? *Describe the Yellowstone, Platte, and Kansas rivers.*

WESTERN TERRITORY,

OR,

TERRITORY OF OREGON.

Q. How is this Territory *situated ?*

A. Between the Rocky mountains and the Pacific ocean.

Q. Which is the principal *river ?*

A. The Columbia, or Oregon.

1. The *Columbia*, 2d *class*, *rises* west of the Rocky mountains, and *running* in a southwest direction, *falls* into the Pacific ocean.

2. The *principal tributaries* of the Columbia, are the Multnomah, Lewis' river, and Clarke's river, all of the 3d *class*, and all of which *join* the Columbia on the S. E. side ; the 1st, 125 miles from its mouth ; the 2d, 413, and the 3d, about 600.

Q. Which is the *principal settlement ?*

A. Astoria, belonging to the American Fur Company, 18 miles from the mouth of the Columbia.

Questions.—*Map of the World.*

Where is the Western Territory situated ? *Describe Columbia, Multnomah, Lewis', and Clark's rivers.* Where is Astoria ?

ARKANSAS TERRITORY.

Q. How is Arkansas Territory *bounded ?*

A. N. by M... T... and M...; E. by T... and M...; S. by L... and M...; W. by M...

1. The *face of the country* is mountainous in the west, but more level in the east. The *soil* is good.

2. The *climate* corresponds to the neighbouring territories. *Pop.* 14,273.

Q. Which are the principal *rivers ?*

A. The Mississippi, Red, Arkansas, St. Francis, White, and Washita.

The Mississippi is the *eastern boundary;* the Red river the *southwestern.* The *Arkansas* river *rises* in the Rocky mountains, and *flows* into the Mississippi. The *St. Francis* and *White* rivers *both fall* into the Mississippi. The *Washita rises* in this territory, but *passes* into Louisiana.

Q. Which is the *chief town*, and which the *capital ?*

A. A...s is the chief town, A...s the capital.

1. *A...s, small town, is on* the Arkansas river, 65 miles from its mouth. It is an old French settlement.

2. *A...s,* was formerly *called* Little Rock : it *lies* on Arkansas river, 300 miles above the Mississippi.

3. *Dwight* is a Missionary station, in a Cherokee country, near Arkansas river, 130 miles above Arkopolis.

Q. What *Indians* inhabit this Territory ?

A. The Osages and the Cherokees are the principal tribes.

Q. Which are its principal *productions?*

A. Cotton and rice.

Questions.—Map of the United States.

How is Arkansas Territory bounded N. E. S. W. ? *Describe the Red, Arkansas, St. Francis, White, and Washita rivers.* Which is the chief town, and which the capital ? *How are they situated ?*

FLORIDA.

Q. How is Florida *bounded ?*

A. N. by A... and G...; E. by the A...; S. by the G... of M...; W. by the G... of M... and A...

1. The *territory of Florida was held by* Spain, Great Britain, and France, at different times, *until* 1803, *when the last government ceded it* to the United States. Difficulties, however, arising from Spain, the *final surrender* of the whole Territory *was not made until* 1821, and was then *purchased of* Spain, *for* 5 millions of dollars.

2. *The surface* of Florida, like Georgia and other neighbouring states, is low and level; the *soil*, except on the rivers, is barren.

3. The *climate*, which would otherwise be extremely hot in summer, is *refreshed by* breezes from the Atlantic, and the Gulf of Mexico.

4. Its *length* is computed to be 400 miles; medium *breadth*, 140; *square miles*, 45,000. *Pop.* 12,000, the *principal part* of which are Spaniards.

Q. Which is the principal *bay* or *gulf?*

A. The Gulf of Mexico.

Q. Which are the principal *rivers?*

A. The St. Johns, Pirdido, (Per-de'do) and Appalachico'la.

The St. Johns, 5*th class*, is the *largest*; it *rises* in the southern part of the peninsula, and *running* in a northerly direction, *expands into* several lakes, through which its waters *pass*, and *turning* to the east, *empties* into the Atlantic, 30 miles north of St. Augustine. The *Perdido separates* Florida from Alabama. The *Appalachicola*, 3*d class, comes* from Georgia, where it is *formed* by the Chatahoochee and Flint rivers. Its *course* is southerly, and it *falls* into Apalachy bay.

Q. What large *swamp* lies partly in Florida?

A. Okefono'co.

Q. Which is the *chief town*, and which the *capital?*

A. St. A...e is the chief town, and T...e the capital.

Q. What *Indians* live in Florida?

A. The Seminole Indians.

In their late war with the United States, this tribe was greatly diminished, so that, at present, it does not contain more than 6000.

Q. Which are the principal *productions?*

A. Cotton, rice, and sugar.

Questions.—Map of the United States.

How is Florida bounded N. E. S. W.? *Describe the rivers St. Johns, Perdido, and Appalachicola.* Where is Okefonoco Swamp? What cape is at the southern extremity? What islands lie southeast?—Which is the chief town, and which the capital? *How are they situated?*

𝕸𝖊𝖝𝖎𝖈𝖔, 𝖔𝖗 𝕹𝖊𝖜 𝕾𝖕𝖆𝖎𝖓.

Pyramid of Cholula.

Q. How is Mexico *bounded ?*

A. N. by w... T..., M... T..., and A... T...; E. by
M... T..., A... T..., L..., and G... of M...; S. by G...,
and the P... o... ; W. by the P... o...

1. *Mexico was subdued by the Spaniards* under *Cortez,* in
1521, and *until 1821, was a province* of Spain, *governed* by a
viceroy; but at that time, *it declared itself independent.*

2. The *face of the country,* on both the coasts, is low, but
rises *towards the middle,* about 6000 feet, when it *spreads out*
into an extensive plain of 1700 miles in length, *called* table
land.

3. The *soil* of Mexico is generally good, and often fertile.

4. The *climate* varies with the situation. On the *coasts* it
is hot and sickly; but on the *elevated plains,* it is cool, and
healthful.

5. The *length* of Mexico is estimated to be 1820 miles,
medium breadth, 800 ; *square miles,* 1,000,000. *Pop.* in 1820,
8,500,000.

Q. Which are the principal *mountains ?*

A. The range known by the name of the Cordil'.
leras.

This range is a continuation of the Andes in South America, and which, in a *higher latitude, is known by the name of*
the Rocky mountains. Popocat'epetl and Oriza'ba are *volcanic
mountains,* of the 3d class.

Q. Which are the principal *rivers?*

A. The Rio del Norte' and the Colora'do.

1. The *Rio del Norte*, *1st class, rises* in the Rocky mountains, and *flowing* southeasterly, *empties* into the Gulf of Mexico.

2. The *Colora'do*, *2d class, rises* on the west side of the mountains, and *flowing* southwesterly, *empties* itself into the Gulf of California.

Q. Which is the principal *lake ?*

A. Lake Chap'ala.

This lake is 170 *miles west of Mexico*, and is about 90 miles *long*, and 20 *broad*.

Q. What *Gulfs* border on Mexico ?

A. Those of Mexico and California.

Q. Which is the *chief town* and *capital ?*

A. M...o.

1. *M...o*, *3d class*, is *situated* in a delightful valley, which is more than 7000 *feet above the ocean*. It is *near* a beautiful *lake* called Tezcuco, upon whose waters float artificial gardens, in which the people raise vegetables. The *streets of Mexico* are broad, clean, and well lighted. The *public buildings* are magnificent ; and some of them of the most beautiful architecture.

2. The *other principal cities* are Puebla, (Poo-a'bla) *4th class, famous for* its manufactures of earthen ware, iron, and steel, 70 miles S. E. of Mexico; and Guanaxuato, (Gwah-na-kwa'to) *4th class, famous for* its gold and silver mines, 150 miles N. W.

3. Vera Cruz, *6th class*, on the Gulf of Mexico, and Aca-pul'co, *large town*, on the coast of the Pacific, are the *principal sea-ports*.

Q. What *college* does Mexico contain ?

A. A university in Mexico, composed of 150 doctors, in all the departments.

Classical literature receives but little *attention*, but the mathematics, chemistry, natural history, and the fine arts, are very diligently studied. The establishments for the promotion of science in the city of Mexico, are said to be exceedingly splendid.

Q. What is the *religion* of the country ?

A. Roman Catholic.

Q. What can you say of the *minerals* of Mexico ?

A. Mines of gold and silver abound, and are wrought with immense profit.

The *silver mines of Mexico* are *said to yield* ten times as much silver as is obtained from all the mines in Europe.— The two *principal mines* are Guanaxuato. and Zacatecas.—

The *whole annual produce* of these mines averages about 20,000,000 dollars.

Q. What celebrated *curiosity* can you mention ?

A. The *Pyramid of Cholula*, one of the religious monuments of the ancient Mexicans.

This pyramid is situated on the east side of the city of Cholula, 70 miles E. of Mexico. It is four stories *high*, and is *constructed of* unburnt bricks, with layers of clay. The *breadth of its base* is 1423 feet, and it rises to the *height* of 177 feet.

Q. Which are the principal *productions ?*

A. Indian corn is the principal. Cotton and wheat are cultivated, and tropical fruits abound.

Q. What can you say of the *inhabitants ?*

A. They consist of seven classes.—1. *European Spaniards ;* 2. *Creoles*, or whites of European extraction, born in America; 3. *Negroes ;* 4. *Indians ;* 5. *Mestizoes*, or descendants of whites and Indians ; 6. *Mulattoes*, or descendants of whites and negroes ; 7. *Zambos*, or descendants of negroes and Indians.

Q. What is the *government* of Mexico ?

A. The country formerly belonged to Spain, but recently the inhabitants have declared themselves independent, and have established a *republican government.*

Q. What is the *character* of the native Mexicans?

A. They are intelligent, indolent, and quick of apprehension; but much addicted to strong liquors.

Questions.—Map of the World.

How is Mexico bounded N. E. S. W. ? What range of mountains do you notice ? What is this range called in the U. States ? Which side of these mountains does the Colorado rise ? *What is its direction ? Where does it empty ? Describe the Rio del Norte.* In which part of Mexico is lake Chapala? What gulf is in the west ? Where is cape St. Lucas ? Where cape Corrientes ? What is the chief city ? *Where is Santa Fe ? Acapulco ? Vera Cruz ?*

Guatemala,
OR REPUBLIC OF CENTRAL AMERICA.

Q. How is Guatemala (Gwah-te-mah'la) *bounded?*

A. N. by m... and bay of h...; E. by the c...

s... and i... of d...; S. by the p...; W. by p..
and m...

1. *Guatemala*, or the Republic of Central America, *contains* five *independent states;* Guatemala, Salvador, Honduras, Nicaragua, and Costa Rica; united together in one republic, under a general congress. *The independence of the country was declared* in 1821, but was not finally effected till 1823. The *constitution was modelled after* that of the United States.

2. The *face of the country* is much diversified, being low and swampy on the *coast*, and very mountainous in the interior.

3. The *soil* is in general extremely fertile.

4. The *climate*, like all southern countries, is hot, and on the coast is sickly.

5. *Square miles*, 300,000 ; *Pop.* 1,500,000. The *inhabitants* are generally Indians, and little is known about them.

Q. Which are the principal *mountains?*

A. A great range which connects the Andes of South America and the Cordilleras of Mexico ; many peaks of which are volcanic.

Q. Which is the principal *river?*

A. The St. Juan'.

The St. Juan, 6th class, *forms* the *outlet of* lake Nicaragua, *flowing* easterly into the Carribbean sea.

Q. Which is the principal *bay?*

A. The bay of Hondu'ras.

Q. Which is the principal *lake?*

A. Nicaragua, (Nic'ar-aw'gua.)

This lake is about 300 *miles in circumference*, and *communicates with* the Carribbean sea by the river St. Juan. *The project has been suggested* of opening a communication between the Atlantic and Pacific oceans, by means of a canal, from the lake Nicaragua to the Pacific. This, in time, will probably be effected.

Q. Which is the *chief town* and *capital?*

A. G...a

1. *G...a, 4th class, lies* near the Pacific ocean. It was *founded* in 1524, but was *overwhelmed* in 1751 by an earthquake, and by matter from a volcano. The city was, however, *rebuilt* on the same spot, and *again destroyed* in 1775, with nearly all its inhabitants. The *present city stands* 25 miles south of the old town, and is magnificent.

2. The *other principal town* is Leon, *6th class,* the chief town of the province of Nicaragua.

Q. Which are the principal *productions?*

A. Grain, cotton, wool, and dye-woods; particularly log-wood and mahogany.

Q. What is the *government?*

A. A confederated republic.

Questions.—Map of the World.

How is Guatemala bounded N. E. S. W. ? What lake does Guatemala contain, *and where does it empty?* Where is cape Blanco ? Where is the bay of Honduras ? What is the capital? *How situated?*

British America.

Quebec.

Q. What *possessions* have the *British* in North America ?

A. The Island of Newfoundland, and the provinces of 1. Nova Scotia; 2. New Brunswick; 3. Upper Canada; 4. Lower Canada; besides which, they claim New Britain.

Q. How are these possessions *bounded?*

A. N. by f... o...; E. by the a...; S. by the u... s...; W. by the p... o... and r... s...

Q. What is the *character* of the inhabitants of British America ?

A. They consist of French, English, and Americans: the *French* are ignorant and superstitious; the *upper classes* of the *English* and *Americans* are superior in character and information.

Questions.—Map of the World.

How are the British possessions bounded N. E. S. W. ? What

bay do they include ? *What divisions are made round Hudson's bay ?* What lakes lie partly in this country ? What rivers empty into Hudson's bay ? What strait connects Hudson's bay and the Atlantic ? What is the south end of Hudson's bay called ?

NEWFOUNDLAND.

Q. How is Newfoundland' *situated ?*

A. It is separated from Labrador by the strait of Bellisle.

1. The *face of the country* is hilly ; the *soil* barren.

2. The *climate* is severely cold, but healthful.

3. The *length* is 380 miles; *breadth*, from 40 to 280 ; the *population* varies according to the prosperity of the fisheries; it sometimes amounts to 70,000, but does not generally exceed 50,000.

Q. Which is the *chief town* and *capital ?*

A. St. J...n's.

1. *St. J...n's, 6th class, lies* on the east coast, on a bay of the same name. It has the best *harbour* in the island. *In* 1816 and 1817, it was nearly *destroyed by fire.*

2. The *other principal towns* are Placentia and Bonavista, *small towns.*

Q. Which is the prevailing *religion ?*

A. The Roman Catholic.

Q. What gives Newfoundland its *importance ?*

A. Its Cod Fisheries.

The *chief places where the cod fish are taken are* the Grand Bank and Green Bank. The *former lies* 100 miles southeast of the island, and is 300 miles *long*, and 75 *broad ;* the *latter* is east of the Grand Bank, and is 240 miles *long*, and 120 *broad. These fisheries employ* about 3000 *vessels* and 100,000 *seamen belonging to* Great Britain, France, and the United States.

Q. What is the *government ?*

A. It is under the government of Great Britain, administered by an admiral.

Questions.—Map of the World.

How is Newfoundland situated ? What strait separates it from Labrador ? Which way is the Grand Bank from Newfoundland ? What is the capital ? *How situated ?*

NOVA SCOTIA.

Q. How is Nova Scotia *bounded ?*

A. N. by the B... of F..., N... B..., and G... of St. L... ; E. and S. by the A... ; W. by the A..., and B... of F...

1. *Nova Scotia originally belonged* to the French, who *called it* Acadia, but few settlements were made till 1749, when 3000 English colonists came over and planted themselves at Halifax.

2. The *climate* is cold, but healthful; much of the *soil* is fertile.

3. *Length,* 250 miles ; *breadth,* from 30 to 60; *squars miles,* 15,000. *Pop.* 100,000, *principally from* New England. After these the Scotch and Irish are most numerous.

Q. What *bay* lies near Nova Scotia ?

A. The Bay of Fundy, between Nova Scotia and New Brunswick.

The Bay of Fundy is remarkable for its tides, which, in some parts, rise 50 or 60 feet, and with so much *rapidity,* that cattle are sometimes overtaken by the waters, and drowned.

Q. What *cape* deserves notice ?

A. Cape Sable, on the southern extremity of the province.

Q. Which is the *chief town* and *capital ?*

A. H...x.

1. *H...x, 6th class, stands on* Chebucto Bay, in the centre of Nova Scotia. Its *harbour* is good, and is the principal naval station of Great Britain, in North America.

2. The *other principal towns,* are Liverpool and Pictou, *small towns.*

Q. What *island* lies near Nova Scotia ?

A. St. John's, or Prince Edward's.

This island is 100 miles *long.* The *principal town* is Charlotteville, *small town.*

Q. Which are the principal *exports* of Nova Scotia ?

A. Fish, Lumber, and Plaster of Paris.

Q. What is the *government ?*

A. It is under a lieutenant governor, as are the other provinces of British America; all, however, subject to a governor general, who resides at Quebec.

Questions.—Map of the United States.

How is Nova Scotia bounded N. E: S. W. ? What cape lies at the south ? What is the capital ? *How situated ?*

NEW BRUNSWICK.

Q. How is New Brunswick *bounded ?*

A. N. by L... c...; E. by g... of St. l.... and n...
s...; S. by the b... of f...; W. by m...

1. The *climate* is cold, but healthful; the *soil* is generally productive.

2. *Length,* 200 miles; *breadth,* 160; *square miles,* 30,000. *Pop.* 100,000.

Q. Which is the principal *river?*

A. St. John's.

St. John's river, 5th class, rises in Maine, and *flowing* northerly, easterly, and southeasterly, *empties* into the Bay of Fundy. It is *navigable* for sloops 80 miles, and for boats 200.

Q. Which is the *chief town,* and which the *capital?*

A. St. J...n's is the chief town; F...n the capital.

St. J...n's, large town, is *situated* near the mouth of St. John's river. *F...n, small town, lies* on the same river, 80 miles from its mouth.

2. *St. Andrew's, small town,* on an arm of Passamaquoddy bay, is the *other most considerable town.*

Q. Which are the principal *productions* and *exports?*

A. The principal productions are grass and grain; the principal exports are timber and fish.

Questions.—*Map of the United States.*

How is New Brunswick bounded N. E. S. W.? What river do you notice? *Where does it rise? Where does it empty?* Which is the chief town? Which the capital? *How situated?*

UPPER CANADA.

Q. How is Upper Canada *bounded?*

A. N. by N... B... and L... C...; E. by L... C... and the U... S...; S. and W. by the U... S...; from which it is separated by Lakes O..., E..., H..., and S...

1. The *face of the country* on the St. Lawrence, and the lakes, is generally level, and contains a rich *soil.*

2. The *climate* is cold, though warmer than that of Lower Canada.

3. *Length,* and *breadth,* unknown: *square miles,* estimated at 290,000. *Pop.* about 150,000.

Q. Which are the principal *rivers?*

A. The St. Lawrence, Ou'tawas, and Niag'ara.

The *St. Lawrence, 1st class,* is the *outlet* of the great northern lakes. Its *course* is northeast, and its *place of discharge,* the Gulf of St Lawrence. The *Outawas,* or Ottawa, *3d class, rises* on the N. side of Lake Huron, and *flowing* generally southeasterly, *divides* Upper and Lower Canada, and *falls* into the St. Lawrence. *Niagara, 6th class, connects* Lake Erie and Ontario.

Q. Which are the principal *lakes ?*

A. One half of the lakes Superior, Huron, Erie, Ontario, St. Clair, Rainy Lake, and Lake of the Woods, is included in Upper Canada.

Q. What *bay* lies in Upper Canada ?

A. The bay of Quinti.

The *bay of Quinti* is 70 miles *long*, and from 1 to 6 *broad.* It *lies* at the northeast end of lake Ontario.

Q. Which is the *chief town* and *capital ?*

A. Y...k.

1. *Y...k, small town, lies* on the northwest side of Lake Ontario, and has a beautiful and commodious *harbour.*

2. The *other principal towns,* are Newark, Queenstown, and Kingston, *small towns.*

Q. Which are the principal *productions* and *exports ?*

A. Wheat, Indian corn, pot ashes, and lumber.

Questions.—Map of the United States.

How is U. C. bounded N. E. S. W. ? What river separates U from L. Canada ? What lakes lie on the southern border ? What lake lies between Huron and Erie ? Between what lakes are the falls of Niagara ? What is the chief town and capital ? *How situated ?*

LOWER CANADA.

Q. How is Lower Canada *bounded ?*

A. N. by n... b... ; E. by n... b..., and g... of St. l... ; S. by n... b..., u... s..., and u... c... ; W. by u... c..., and n... b...

1. Canada was originally *settled* by the French, by whom it was *held* till 1759, when Gen. Wolfe *took* it, upon which it was surrendered to the English, in whose power it has since remained.

2. The *face of the country* is hilly, but the *soil* is fertile.

3. The *climate* is intensely cold in winter. The *summers* are short, but warm.

4. *Length,* and *breadth,* uncertain ; *square miles,* estimated at 280,000. *Pop.* 364,000.

Q. Which are the principal *rivers ?*

A. The St. Lawrence and the Ou'tawas.

The *St. Lawrence, 6th class, passes* through this province ; the *Outawas, 3d class, empties* into the St. Lawrence, near Montreal.

The *other rivers* are the Sorelle', St. Francis, and Chaudiere, (Shaw-di-are'.)

Q. Which is the *chief town,* and which the *capital ?*

A. M...l is the chief town, and Q...c the capital.

8*

1. *M...l, 6th class, stands* on the St. Lawrence, 180 miles above Quebec. *Its commerce* is extensive.

2. *Q...c, 6th class, stands* on the St. Lawrence, 350 miles from its mouth. It is strongly *fortified* both by nature and art.

Q. Which is the prevailing *religion?*

A. The Roman Catholic.

Q. What *Island* is attached to this province?

A. The Island of Cape Breton, (Bre-toon'.)

Formerly fishery *was extensively carried on* from this island, but *at present its inhabitants are employed* in their coal mines. *Pop.* 3000.

Q. Which are the principal *exports?*

A. Furs, lumber, and pot ashes.

Questions.—*Map of the United States and World.*

How is Lower Canada bounded N. E. S. W.? *Describe the rivers.* What gulf does L. C. contain? What islands are in this gulf? Which is the chief town, and which the capital? *How situated?*

NEW BRITAIN.

Q. How is New Britain *bounded?*

A. N. by the F... O...; E. by the A...; S. by C... and the U... S...; W. by the P... O... and R... S...

1. This is a vast country *divided* by Hudson's bay into two parts—the *eastern* and the *western.* The *eastern comprehends* Labrador and East Maine; the *western,* New South and New North Wales.

2. It is *inhabited* chiefly by Indians, and visited only by such as are in quest of furs.

3. The *climate* is extremely cold, and the *soil* so poor as to bear only shrubs and moss.

Q. Which are the principal *rivers?*

A. Macken'zie's and Nelson's.

Mackenzie's river, 1st class, *forms* the outlet of Slave Lake. After a winding *course,* it *falls* into the Frozen ocean. *Nelson river,* 2d class, is the *outlet* of lake Win'nepeg; its *course* is northeast, and it *falls* into Hudson's Bay.

Q. Which are the principal *bays?*

A. Baffin's bay and Hudson's bay. The southern part of Hudson's bay, is called James' bay.

Q. What can you say of the *lakes?*

A. The country abounds with small lakes.

Q. What *trade* is carried on in these regions?

A. The fur trade.

This trade is carried on by two companies, the *Hudson Bay Company,* and the *North West Company.* *The latter company* extend their trade *to* the Rocky mountains, and even to the Frozen Ocean. *The men employed, travel* in birch bark canoes. In these they proceed thousands of miles into the interior, carrying their canoes on their shoulders, when the rivers are interrupted by falls, or when they would pass from one river to another.

Q. What *Indians* inhabit this country ?

A. The Esquimaux, (Es'ke-mo) and other tribes

The *Esquimaux inhabit* the coast of Labrador, and the shores of the Frozen ocean. The *interior is inhabited* by tribes of which little is known.

Q. What is the *character* of the Esquimaux ?

A. They are dwarfish, indolent, dull, and filthy ; but mild and tractable.

Questions.—*Map of the World.*

How is New Britain bounded N. E. S. W. ? *What are the divisions?* What bay does N. B. include ? What is the S. part of Hudson's bay called ? *Describe the rivers of New Britain.*

Russian Possessions.

Q. Where are the *Russian possessions* in North America ?

A. They are on the North West Coast, extending from Behring's strait to Portlock harbour, seven degrees, or about 480 miles from north to south.

Little is known of this country. The *white inhabitants are estimated at* 1000. The *Indians subject to Russia, at* 50,000. The *principal employment of both whites* and *Indians,* is hunting and fishing.

Questions.—*Map of N. America.*

Where are the Russian possessions in North America situated ? What is the most western cape of N. America ? What peninsula south of it ? What sound near Alaska ? Where is Portlock Harbour ?

Greenland.

Q. To what country does *Greenland belong ?*

A. To Denmark.

Q. How is it *bounded?*

A. N. by regions unknown ; E. and S. by the A... ; W. by D... strait, and B... B...

1. Greenland, until lately, was *supposed to be united to* the continent of America, but recent discoveries render it probable that it is not.

2. The *face of the country* is dreary, on account of naked mountains and perpetual ice.

3. It *produces* only a few miserable trees and shrubs.

4. *Length* and *breadth* uncertain. *Pop.* 15,000.

Q. What *cape* lies at the southern extremity ?

A. Cape Farewell.

Q. What is the *religion* of the Greenlanders ?

A. Formerly they were Pagans, but they have been converted to Christianity by Moravian missionaries.

Q. Which are the principal *missionary settlements* ?

A. New Hernhut, Lichtenfels, and Lichtenau.

Q. What can you say of the *Greenlanders* ?

A. They are small of stature ; indolent, weak, and extremely filthy in their mode of living.

Q. What *animals* are found in Greenland ?

A. The reindeer, foxes, hares, bears, and seals.

Questions.—Map of the World.

How is Greenland bounded N. E. S. W. ? What is the southern cape ? What strait separates it from Labrador ? What island lies N. E. from Greenland ?

West Indies.

Negroes Dancing.

Q. Where are the West India islands *situated* ?

A. On the eastern side of the American continent, between North and South America.

Q. Which are the *extreme islands* towards the several points of the compass ?

A. Trinidad' is the most southern ; Barba'does the most eastern ; Cu'ba the most western ; and the Baha'ma islands the most northern.

Q. How are the West India islands commonly *divided ?*

A. Into Great An'tilles, Little An'tilles, Baha'mas, and Car'ribbee islands.

1. The *Grand Antilles* are the four largest islands—Cuba, Hispaniola, Jamaica, and Porto Rico

2. The *Little Antilles* are Curracoa, Buen Aire, and some others, near the coast of South America.

3. The *Bahamas* are those islands north of Cuba and Hispaniola. These islands are said to be about 500 in *number*, of which not more than 12 are of any considerable extent, and are nearly all destitute of inhabitants.

4. The *Carrib-be'an islands* are Trinidad, and all north of it, till you come to Porto Rico.

5. The *Virgin* islands are included among the Carribbean islands.

6. The *Bermuda* islands are generally spoken of in connexion with the West Indies. They are 600 miles *east of Carolina*, and are about 300 in *number*, with a *population* of 10,000.

Q. How are the *Car'ribbee islands divided ?*

A. Into Leeward and Windward.

Tabular View of the principal West India Islands

	Names.	Belonging to.	Sq. m.	Pop.	Chief Towns.
Gt. Antilles. Bahamas	Bahama Providence Abaco, &c.	Great Britain	5,500	15,000	Nassau.
	St. Domingo or Hayti	Independent	28,000	650,000	Cape Henry
	Cuba	Spain	50,000	620,000	Havannah.
	Jamaica	Britain	6,400	360,000	Kingston.
	Porto Rico*	Spain	4,000	130,000	St. Juan

Leeward Islands.

Carribbean Islands	Guadaloupe†	France	675	115,000	Basse Terre.§
	Antigua‡	Britain	93	36,000	St. Johns.
	St. Croix§	Denmark	100	31,000	Christianstadt.
	St. Christophers	Britain	70	28,000	Basse Terre.
	Dominica	do.		26,000	Rosean.‖
	St. Eus'atia	Netherlands	22	20,000	The Bay.
	Mariegalante	France	90	12,000	Basse Terre.
	Montserat	Britain	78	11,000	Plymouth.
	Tortola	do.	90	11,000	Road Harbour
	Nevis	do.	20	10,000	Charlestown

Names	Belonging to.	Sq. m.	Pop	Chief Towns.
St. Bartholomew	Sweden	60	8,000	Gustavia.
Virgin Gorda	Britain	80	8,000	
St. Martin	Netherlands	90	6,000	
St. Thomas	Denmark	40	5,000	
Anguilla	Britain	30	800	
Windward Islands.				
Martinique¶	France	370	95,000	St. Pierre.††
Barbadoes	Britain	166	93,000	Bridgetown.
Grenada	do.	110	31,000	St. George.
Trinidad	do.	1,700	25,000	Port of Spain.
St. Vincent	do.	130	23,000	Kingston.
St. Lucia	do.	225	16,000	Carenage.
Tobago	do.	140	16,000	Scarborough.
Margaritta	Colombia	350	14,000	Ascension.
Curracoa‡§	Netherlands	600	8,500	Williamstadt.

(left margin: Caribbean Islands.)

* Porto Re-ko. † Gaud-a-loop. § Bass-Tare. ‡ An-te-ga. § San-ta-Cruz.
|| Ro-so. ¶ Mar-tee-neek. †† Saint-Pe-aer. ‡§ Cur-a-so.

Q. What is the *climate* of the West India islands?

A. In summer the climate is hot and unhealthful, especially to strangers; but in winter the temperature is delightful.

Hurricanes occur in autumn, which sometimes cause the greatest destruction. Earthquakes and violent thunder and lightning are also common.

Q. What is the *soil?*

A. It is in general very fertile.

Q. Which are the principal *productions?*

A. Sugar, rum, and molasses.

The *other productions* are cotton, indigo, coffee, cocoa, ginger, &c. *Of fruits, the principal are* oranges, lemons, limes, pine apples, and figs.

Q. Which are some of the principal *towns* in these islands?

A. *Havan'nah* is the chief town of Cuba—*Kings'-ton* of Jamai'ca—*St. Juan'* of Porto Rico, (Por'-to-Re'ko)—*St. Pierre*, (Saint-Pe-aer) of Martinique, (Mar-te-neek')—*Cape Henry, St. Domingo*, and *Port au Prince* of St. Domingo.

1. Havannah, *4th class*, is the *largest town in the West Indies*. It is *situated* on the northern coast of Cuba. It has a *harbour* capable of commodiously containing 1000 ships. It is a strongly fortified place. Its *commerce* is immense.

2. Kingston, *5th class*, is the *chief town of Jamaica*, and indeed of the British West Indies. It has an excellent *harbour*, and an extensive *commerce.*

3. St. Pierre, *5th class*, in Martinique, is the *principal town of the French West Indies*, and enjoys an extensive *commerce*

Q. What is the *population* of all the West India Islands ?

A. About 2,400,000 ; three fourths of whom are blacks, and chiefly slaves.

Q. What is the *character* of the inhabitants ?

A. It is very various. The *white* inhabitants of the British West India Islands are said to be hospitable and generous, but high minded and contentious; the *creoles* are indolent, and fond of pleasure, but generous, high minded, and frank.

Questions.—Map of the World.

Where are the West India Islands situated ? Which is the most northern, eastern, southern, western ? *Between what latitudes do they lie ?* Which is the largest ? Which the 2d, 3d, 4th ? Which way is Florida from the Bahama Islands ? Which way from the Bahamas is Cuba ? Which way from Cuba is Jamaica ? What sea lies between these islands and S. America ? Which way is St. Domingo from Cuba ? Where are the Bermudas ?

SOUTH AMERICA.

South American Indians.

Q. How is South America *bounded ?*

A. N. by the c... s... ; E. by the a... ; S. it terminates in c... h...; W. by the p... o...

1. *Greatest length,* 4570 ; *greatest breadth,* 3230. It contains 6 or 7,000,000 *square miles,* and from 10,000,000 to 15,000,000 of *inhabitants,* of whom nearly one half are independent Indians.

2. The *climate* of South America embraces almost every extreme of heat and cold; but the temperature is generally higher than in the same latitude on the eastern continent.

Q. Which are the principal *mountains?*

A. The Andes.

1. The *Andes run* the whole length of South America, *at a distance of* from 50 to 200 miles *from the shores of the Pacific.*

2. The *height of the Andes varies* from 1000 feet above the level of the ocean, to more than 21,000 feet. Many peaks are volcanic.

3. From the great range, many *branches* run off towards the east. The *principal branch* or chain, is that of *Venezuela,* (Ven-e-zwa'la) which stretches along the northern coast of South America, towards the island of Trinidad.

Q. Which is the most important *island?*

A. Terra del Fu'ego.

1. *Terra del Fuego lies* on the south of South America, and is *separated from the continent* by the strait of Magel'lan.

2. Besides this are the Falkland *islands,* N. E. of Terra del Fu'ego; and the *island* of Juan Fernan'dez, on the coast of Chili, *celebrated* as the residence of Alexander Selkirk, a sailor, who lived there four years in solitude, which circumstance was the foundation of the interesting story of Robinson Crusoe.

Q. Which are the principal *rivers?*

A. The greatest are the Am'azon, the La Pla'ta, and the Orono'co.

1. The *Amazon,* 1st *class, rises* in the Andes, and *flowing* first northerly, and then northeasterly, *empties itself* into the Atlantic, under the equator, by a mouth, 180 miles wide.

2. The *La Plata, taken with the Paraguay,* 1st *class,* is *formed* by the union of the Uruguay and Parana, both of which *rise* in Brazil; it *flows* into the Atlantic.

3. The *Oronoco,* 2d *class, rises* in lat 5° N. Its *course* is very crooked, somewhat resembling the figure 6. It *empties* itself on the northern coast, opposite the island of Trinidad.

4. The *other principal rivers* are (1.) the Uruguay, 2d *class,* which is the *eastern* branch of the Plata, and *flows* southwesterly: (2.) the Parana, 2d *class,* the western branch of the Plata, *flowing* southwesterly; (3.) the Paraguay, 2d *class,* which *rises* near a ridge of the Andes, and *flowing* southerly, *joins* the Parana. (4.) The Madeira, 2d *class,* which *rises* in the Andes in Buenos Ayres, and *flowing* northeasterly through Brazil, *empties* into the Amazon. (5.) The Tocantins, 2nd *class,* which *rises* between the parallels

of 18° and 19° S. lat. and *flowing* northerly, *empties* itself into the Amazon. (6.) The Magdalena, *2d class*, which *rises* among the Andes, and *flowing* northerly *falls* into the Carribbean sea.

Q. What can you say of South America in respect to its *productions?*

A. It is distinguished for its productions, mineral, animal, and vegetable.

1. It is *particularly distinguished for* its mines of gold and silver. It *yields platina*, the heaviest of the metals, and also diamonds. Jesuits' bark *is peculiar* to South America.

2. Some of the *animals which are remarkable*, are the jaguar, or American tiger, which is very ferocious; the llama, or Peruvian camel, and the vicuna, or Peruvian sheep; both are *used*, but particularly the llama, as beasts of burden.

3. Among the *birds which inhabit the Andes*, is the Condor, the *largest* bird known that flies. It *soars* to the height of 20,000 feet. Its *wings*, when extended, *measure* from 12 to 16 feet, and the greater quills 2 feet 9 inches long. It *preys* upon calves, sheep, and goats, and has been known to carry off children 10 years of age.

Q. What is the *religion* of the South Americans?

A. The Roman Catholic, except of the independent Indians, and the possessions of the English and Dutch.

Q. What is the *character* of the South American Indians?

A. They resemble, in many respects, the North American Indians.

Q. What are the *divisions* of South America?

A. The Republic of Colom'bia—Guiana, (Gee-an'na)—Peru'—Bolivia—Brazil, (Bra-zeel')—Buenos Ayres, (Bo-nos-Ai'res) or the United Provinces of La Pla'ta—Chili, (Chee'le) and Patago'nia.

N. B. South America has been, until lately, under the dominion of Spain and Portugal, but it is all liberated and under republican governments, except *Guiana, Brazil, and Patagonia*. The state of these new republican governments is rather unsettled, and further changes may be expected.

Questions.—Map of S. America.

How is South America bounded N. E. S. W.? How is it divided? What countries N. of the equator? What S.? What is the S. cape of S. America? What is the most N.? Which are the most E. and W. capes? Where is cape North? Where C. St. Antonio?

Where St. Maria ? Where is the B. of All Saints ? What Strait between Patagonia and Terra del Fuego ? What is the great chain of mountains in S. America ? What is their course ? On which coast do they run ? Where is Mt. Chimborazo ? What is the largest river of S. America ? *Where does it rise ? What is its course ? Where does it empty ?* Which are its principal tributaries from the south ? *What is their general course ?* What are the rivers of Guiana ? Where is the Oronoco ? Where is St. Francis, and where does it empty ? Where is the Paraguay ? What are its branches ? Where is the Magdalena ? What is its principal branch ? Why are there no large rivers on the W. coast of S. America ? What Islands lie between S. and N. America ? What Islands lie N. of the mouth of the river Oronoco ? On what Island is Cape Horn ? What Islands lie N. E. of it ? Where is the Island of Chiloe ? Of Juan Fernandez ? Islands of Galapagos ? *What is the latitude of Quito ? Of Lima ? Of St. Jago ? Of Cape Horn ? Of Buenos Ayres ? Rio Janeiro ? Mouth of the Amazon ?* Where is the desert of Ata cama ? Where is lake Titicaca ?

Republic of Colombia.

Andes near Quito.

Q. How is Colombia *bounded*?

A. N. by the C... S...; E. by the A... and G...; S. by G..., B..., and P...; W. by the P... O... and G...

1. *This republic is composed* of the two provinces of New Grana'da and Venezue'la, both of which, until within a few years, were *styled* Terra Firma, and were *subject* to Spain.

2. The *face of the country*, in general, is mountainous, particularly the northern and western parts. The *soil is* fertile.

3. The *climate* on the *sea coast* is hot and unhealthful; in

the *mountainous regions* the climate varies with the elevation. On the highest summits, winter reigns perpetually.

4. *Length,* from E. to W. about 1320 miles; *breadth,* from N. to S. about 1080; *square miles,* 1,100,000. *Pop.* 2,600,000.

Q. Which are the principal *mountains?*

A. The Andes.

1. The *Andes come from* Peru on the south, and *pass* through the country. In their progress, however, they *divide into* three *branches,* the western is the *great* range of the Andes; the *eastern* is the Venezuela chain.

2. Mount Chimbora'zo, *2d class,* near the city of Quito, is the *highest summit,* and indeed the highest mountain in America, being 21,440 feet above the level of the ocean.

3. Cotopax'i, *3d class,* is about 40 miles S. E. from Quito, (Kee'to) and is *the highest volcano in the world.* Its explosions are frequent, and tremendous.

Q. What *lake* does Colombia contain?

A. Maracay'bo lake.

This lake is 180 miles *long,* and 100 *broad.* It *empties* into the gulf of Maracay'bo.

Q. What *bays* can you mention?

A. The principal are the Bay or Gulf of Da'rien, of Panama', and of Guaya⟨quil⟩, (Gwai-a-kill'.)

Q. Which is the principal *river?*

A. The Oronoco.

1. The *Oronoco, 2d class, rises* in lat. 5° N. and *flowing* circuitously, *empties* itself into the ocean near the island of Trinidad by 50 *mouths,* the *two most distant* of which are 180 miles *apart.* It is *navigable* 740 miles from the ocean.

2. The *Meta, 3d class,* is a branch of the Oronoco; it *rises* at the foot of the Andes, and is *navigable* 370 miles.

3. The *Magdale'na, 2d class, rises* among the Andes, and *flowing* northerly, *empties* itself into the Carribbean sea. It is *navigable* 600 miles. The Cauca, *3d class,* is the great *western branch of the Magdale'na;* it *rises* among the Andes and *runs* northerly.

Q. Which is the *chief town* and *capital?*

A. Q...o.

1. Q...o, *4th class,* is *situated* among the Andes, under the equator. It is 9000 feet *above the level of the ocean,* but enjoys a mild and healthful *climate.*

2. *Santa Fe',* or *Santa Fe' de Bogota, 5th class,* was formerly the seat of government of New Granada. It is *situated* on a

branch of the Magdale'na, in a fertile plain, elevated nearly 8700 feet above the level of the sea. Popayan, *5th class*, the seat of the mint, is 200 miles N. E. of Quito.

3. The *principal town* near the *northern coast* is *Caraccas*, *5th class*, which was nearly destroyed in 1812, by an earthquake, which buried 12,000 persons in ruins.

4. On the *Carribbean sea* are Cumana, Maracay'bo, and Çarthage'na, all of the *4th class*.

5. On the *Pacific are the ports* of Panama' and Guayaquil', *large towns.*

Q. What *island* belongs to Colombia ?

A. Margaritta.

Q. What is the prevailing *religion?*

A. Roman Catholic.

Q. Which are the principal *productions?*

A. Grain, cotton, indigo, sugar, and tobacco.

Q. What is the *government?*

A. Republican.

Questions.—*Map of S. America.*

How is Colombia bounded N. E. S. W. ? What rivers has it ? *What is their course ? Where do they empty ? What is the latitude of Quito ? Near what mountain does it lie? Which way is Santa Fe de Bogota from Quito ? What ports on the I. of Darien ? Where is Lake Maracaybo ?* Which is the chief town and capital ?

Guiana.

Anaconda.

Q. How is Guiana (Gee-an'na) *bounded?*

A. N. by the R... of C... and the A... ; E. by the A... ; S. by the river A... which separates it from B...; W by the R... of C...

Q. To whom does Guiana belong ?

A. To the Colombians, English, Dutch, French, and Portuguese.

1. The *boundaries* between these divisions are not yet determined.

2. *English Guiana* is sometimes *called* Demera'ra; *Dutch Guiana*, Surinam'; *French Guiana*, Cayenne.

3. The *face of the country* is level and low *on the coast*, but in the *interior* is more elevated. The *soil* is fertile.

4. The *climate* is unhealthful *near the ocean*, but in the *interior*, a purer air prevails.

5. *Length*, 500 miles; *breadth*, 400; *square miles*, 160,000. *Pop.* 250,000; not more than 20,000 of which are *whites*; the rest are negro slaves.

Q. Which are the principal *rivers?*

A. Essequebo, (Es-se-kee'bo) Demera'ra, Berbice, (Ber-bees) Surinam', and Maro'ni.

Q. Which are the *chief towns* and *capitals?*

A. Paramar'ibo is the chief town and capital of Dutch Guiana ; Cay'enne of French Guiana ; and Stabro'ek of English Guiana.

Paramaribo, 6th class, is on Surinam river. *Cayenne, small town, lies* on an island. *Stabro'ek, large town, is on* Demera'ra river.

Q. What large *serpent* is found in Guiana ?

A. The Anaconda, or Boa Constrictor, which seizes large and powerful animals for his prey, and crushes them by winding himself around them.

Q. Which are the principal *productions?*

A. Sugar, coffee, cotton, and cocoa.

Questions.—Map of S. America.

How is Guiana bounded N. E. S. W ? What river separates it from Brazil ? Which are the principal rivers of Guiana ? *Where do they empty?* Where is Cape North ? Where is Paramaribo ?—Cayenne ? Stabroek ?

9*

Peru.

Dresses of Gentlemen and Ladies in Peru.

Q. How is Peru *bounded ?*

A. N. by the R... of C...; E. by B... and B... ; S. by B... and the P... o...; and W. by the P... o..

1. The *face of the country* is generally mountainous. but the *soil* of the vallies is fertile.

2. The *climate* on the *mountains* is always cold, on the *coast* it is hot and unhealthful; while the *intermediate table land,* enjoys a uniform and delightful *temperature.*

3. *Length,* 1000 miles; *breadth,* 500; *square miles,* 500,000. *Pop.* 1,500,000, of which 600,000 are civilized Indians.

Q. Which are the principal *mountains?*

A. The Andes, which pass through the whole length of Peru, parallel with the coast of the Pacific.

Q. Which are the principal *rivers ?*

A. There are no large rivers west of the Andes, the *principal river* in Peru, east of the Andes, is the Ucayle, *3d class,* which *rises* in Bolivia, and *flowing* northerly, *falls* into the Amazon.

Q. Which is the *chief town* and *capital ?*

A. L...a.

1. *L...a, 4th class,* is *situated* on the coast of the Pacific, in the centre of a spacious and delightful valley. It is a place of great *commerce.* Callao (Calyou') seven miles distant, is the *port of Lima* (Lee'ma.)

2. *Cusco, 5th class, is* among the Andes, east of Lima.— *It was formerly* a magnificent city and the residence of the Incas.

3. *Truxillo,* (Tru-keel'yo) *is* 900 miles N. of Lima.

Q. What *minerals* are found in Peru ?

A. Gold, silver, quicksilver, copper, and lead.

There are 70 *gold mines,* and 700 *silver mines,* and four of *quicksilver.* The *annual produce of these mines* is estimated at 4,000,000 dollars.

Q. What *lake* lies partly in Peru ?

A. Lake Titicaca.

This lake *borders* on Bolivia. It is 240 miles in *circumference. Upon an island, in this lake, lived* Manco Capac, the first of the Incas, and the founder of the Peruvian monarchy.

Q. What is the *religion ?*

A. Roman Catholic.

Q. Which are the principal *productions ?*

A. Grain, cotton, sugar, fruits, and Jesuits', or Peruvian bark.

Q. What is the *government ?*

A. Republican.

Questions.—Map of S. America.

How is Peru bounded N. E. S. W. ? What mountains pass through it ? Which is the chief town and capital ? *Where situated ? What seaport lies near to Lima ? What town lies near the sea N. of Lima? Where is Cusco, the ancient capital of the Indians ?*

Republic of Bolivia, or Bolivar.

Q. How is Bolivia *bounded?*

A. N. by P... and B...; E. by B...; S. by the U... P...; and W. by the P... O... and P...

1. The republic of Bolivia is *composed of* the country heretofore called Upper Peru. It *consists of* seven separate territorial governments, now called departments; these are, beginning in the south, Potosi, Charcas, or La Plata, Cochabamba, La Paz, Santa Cruz, or Puno, Moxos, and Chiquitos; all of which were formerly annexed to Buenos Ayres. It is *named after* Bolivar, the distinguished individual, who has exerted himself with so much success for South American liberty, and was instrumental in the emancipation of this country. *The independence was declared* on the 6th of July, 1825.

2. The *face of the country* is generally mountainous, and steril; the *soil* of the vallies, however, is exuberant.

3. The *climate* of the *mountainous* regions is cold; that of the *table* or *level land* is milder, and very salubrious.

4. Probable *length*, 700 miles; probable *breadth*, 700; *square miles*, 450,000. *Pop.* estimated at 1,740,000, of which number 1,155,000 are *Indians*.

Q. Which are the principal *mountains?*

A. The Andes, which consist of two ranges, the eastern and western, both of which pass through the country from south to north.

The *eastern* range is lofty, and its summit is covered with perpetual snow. The *western* range is lower, and more irregular.

Q. Which are the principal *rivers?*

A. The Madeira, Ucayle, and Pilcomayo.

1. The *Madeira, 2nd class, rises* in the Andes, in the department of Chiquitos, in several branches, which *unite* near the line between Bolivia and Brazil, and *flowing* northeast *falls* into the Amazon.

2. The *Ucayle, 3d class,* is *formed by the union of* several rivers, some of which take their *rise* in a small lake, which forms part of the northern boundary of Bolivia. The river *flows* northerly into Peru, and *falls* into the Amazon.

3. The *Pilcomayo, 2d class,* is the largest western branch of the Paraguay; it *rises* in about lat. 20 S. and *flows* first easterly, 600 miles, and *then* southeast, 400, when it *falls* into the Paraguay.

Q. Which is the *chief town* and *capital?*

A. P...i.

1. *P...i, 4th class,* is *situated* among the Andes, 11,000 feet *above the level of the ocean*, in the southern part of the republic, on the great post road leading from Buenos Ayres to Lima, 1650 miles from the former, and 1215 from the latter, and about 300 miles to the east of the Pacific.

2. Potosi is only the temporary seat of government. A new capital is to be selected, to be *called* Sucre, after the able general of that name, who so signally contributed to accomplish the independence of the country.

3. The other *principal towns* are La Paz, and Cochabamba, *5th class.*

Q. What *minerals* are found here?

A. Gold, silver, tin, and copper.

The *silver mine* of Potosi is one of the richest in the world. It has already been *worked* three centuries, and in prosperous times is said to *yield* four millions of dollars annually. The *figure* of the mountain which contains the óro is conical, covered with green, red, yellow, and blue spots, which give it an appearance, unlike any mountain in the world. It is entirely bare of trees or shrubs. The *ore is brought down from* the mountain from the *height* of 16,000 feet above the level of the ocean. From lat. 15 to 23 degrees there are 22 silver, and 11 gold mines, which have been worked.

Q. What *desert* lies partly in Bolivia ?

A. The desert of Atacama, *(See U. Provinces.)*

Q. What *lake* lies partly in this country ?

A. Lake Titicaca, *(See Peru.)*

Q. What is the prevailing *religion* ?

A. The Roman Catholic.

Q. What is the *character* of the Indians ?

A. They are sober, honest, and industrious.

Q. Which are the *principal* productions ?

A. Grain, cotton, sugar, Jesuits' bark, and cochineal.

Q. What can you say of its *commerce* ?

A. It is chiefly inland, and is carried on with Peru and Buenos Ayres.

The *trade with Peru has been estimated to amount to seven millions of dollars annually, and that with Buenos Ayres* to eighteen millions.

Q. What is the *government* ?

A. Republican.

A constitution of a republican character, it is expected, will soon be adopted, and a government formed according to its provisions.

Questions.—*Map of S. America.*

How is Bolivia bounded N. E. S. W. ? What rivers rise in this country ? *Which way do they flow ?* What lake lies partly in B. ? What desert ? *In what part ?* Which is the chief town and capital ? *How situated ?*

Brazil.

Slaves searching for Diamonds.

Q. How is Brazil (Bra-zeel') *bounded ?*

A. N. by the R... of C..., G..., and the A... O...,
E. by the A... ; S. by the A..., U... P..., and B... ;
W. by the U... P..., B..., and P...

1. *Brazil* is an immense country, *embracing* more than one
third *of all South America.* The *western part* is known by the
name of *Amazonia.* Its *inhabitants* are Indians.

2. The *face of the country* is remarkably variegated with
mountains, rivers, fertile prairies, and vast and impenetrable
forests.

3. The *climate* of Brazil is rendered pleasant and healthful
by the sea breezes which prevail.

·4· *Length,* 2000 miles ; *greatest breadth,* 2000 ; *square miles,*
2,100,000. *Pop:* 3,000,000, *composed of* whites, negroes, Indians, and mulattoes.

Q. Which are the principal *mountains ?*

A. A range of the Andes, called the Brazilian
Andes, which run along the coast.

Q. Which are the principal *rivers ?*

A. The St. Fran'cis, Tocan'tins, Xin'gu, Tapajos, and Madei'ra.

1. The *St. Francis,* 2d *class, rises* in S. lat. 16°, along the
Brazilian mountains, and *empties itself* into the Atlantic north
of St. Salvador.

2. The *Tocantins,* 2d *class, rises* between the parallels of
18° and 19° S. lat., and *flowing* northerly, *empties* itself into
the Amazon, near the equator.

3. The *Madeira, Xingu,* and *Tapajos,* all of the 2d *class,* rise in the Andes, and *proceeding* from south to north, *fall* into the Amazon.

Q. Which is the principal *cape?*

A. Cape St. Roque.

Q. Which is the *chief town* and *capital?*

A. R...o J...o, or St. S...n.

1. R...o J...o, 3d *class,* is *situated* on the Atlantic coast, about lat. 23°. It is the largest town in South America, and has an extensive *commerce.*

2. *St. Salvador',* 4th *class,* or Bahia, (Bai'ya) 700 miles N. of Rio Janeiro, (Re'o Ja-ne'ro) is a large and rich town. *Pernambuco,* 5th *class,* is 450 miles northeast from St. Salvador: it *carries on* a great cotton *trade.*

Q. Which are the principal *minerals?*

A. Gold and diamonds.

The *gold and diamonds are washed* by slaves out of the sand, which comes down from the mountains. All the great rivers, which flow north into the Amazon, produce gold.— The *principal region in which diamonds are found,* is 400 miles north of Rio Janeiro.

Q. What is the prevailing *religion?*

A. Roman Catholic.

Q. Which are the principal *productions?*

A. Cotton, wheat, sugar, coffee, indigo, and cochineal.

Q. Which are the principal *exports?*

A. Cotton, sugar, coffee, hides, tallow, gold, and diamonds.

Q. What is the *government?*

A. Monarchical and despotic.

N. B. The present ruler of Brazil, who is now styled Emperor, was King of Portugal; he removed the court from Lisbon in Portugal, and established it at Rio Janeiro, in Brazil. He has lately abdicated the throne of Portugal in favour of his daughter of 10 years of age, who is betrothed to her uncle. Of course Brazil is now independent. The emperor has promised the people to secure their rights by a constitution, which, if done, will make the government a *limited monarchy.* (1826.)

Questions.—*Map of S. America.*

How is Brazil bounded N. E. S. W.? What great river lies on the N.? What rivers are tributary to the Amazon? Where is the river St. Francis? *Where does it empty?* What cape is at the eastern extremity of Brazil? What rivers rise in B. which flow into the United Provinces? Which is the chief town and capital?—*How situated? What towns N of this?*

𝕭𝖚𝖊𝖓𝖔𝖘 𝕬𝖞𝖗𝖊𝖘, 𝖔𝖗 𝖀𝖓𝖎𝖙𝖊𝖉 𝕻𝖗𝖔𝖛𝖎𝖓𝖈𝖊𝖘
OF
LA PLATA.

View of Buenos Ayres.

Q. How are the United Provinces of La Plata, or Buenos
Ayres, (Bo-nos-Ai′res) *bounded?*

A. N. by B..., and B... ; E. by B..., and the A... ;
S. by the A..., and P... ; W. by C..., and the P... O...

1. *While under the dominion of Spain, this country was called*
the viceroyalty of Buenos Ayres ; *since the declaration of independence, it is called* the United Provinces of La Plata.

2. The *face of the country north* and *west* is mountainous.
The *southern portion is composed* chiefly of vast plains, called
pampas, which extend 1500 miles in length, and 500 in breadth.
They are covered with high grass, and pasture innumerable
herds of cattle and wild horses.

3. The *climate* is generally temperate and healthful.

4. *Length,* from the northern boundary of Paraguay to St.
George's bay, or gulf, 1529 miles ; *breadth,* on the northern
boundary from Brazil to the Andes, nearly 900 ; on the southern boundary from St. George's bay to the gulf of Guaytecas
300 miles. *Square miles,* probably 850,000. *Pop.* 2,000,000,
of which number 700,000 are civilized Indians.

Q. Which are the principal *mountains?*

A. The Andes, which separate the country from
Chili.

Q. Which is the principal *river?*

A. The La Pla′ta.

1. The *La Plata, with the Paraguay*, 1st *class*, a broad river, *formed* by the union of the Uruguay and Parana, both of which *rise* in Brazil, and after *running*, the former 1200 miles, and the latter 2000, unite above the city of Buenos Ayres.

2. The *Paraguay*, 2d *class, is a branch* of the Parana, and *rises* in Brazil, and *flowing* southerly, joins the Parana at Corrientes. The *Pilcomayo* and *Vermejo*, 2d *class*, are large *branches of* the Paraguay, both of which *rise* in the Andes, the *former flowing* first easterly and then southeasterly, and the *latter* southeasterly, *fall* into the Paraguay.

Q. Which are the principal *capes ?*

A. Cape St. Maria, and cape St. Anto'nio.

Q. What *desert* lies partly in the United Provinces, and partly in Bolivia ?

A. The desert of Atacama.

This desert is 300 miles *long*, and is incapable of *supporting* either animals or vegetables. It *lies* partly in the United Provinces, and partly in the new republic of Bolivia.

Q. Which is the *chief town* and *capital ?*

A. B...s A...s.

1. B...s A...s, 4th *class, is on* the La Plata, 180 miles from the ocean. It enjoys a delightful *climate.*

2. The *other principal towns* are Monte Video, (Mon-te Vee'de-o) 6th *class, on the* La Plata, 90 miles from its mouth. Santa Fe, *large town, situated* at the confluence of the Sala'do with the Para'na.

Q. What is the prevailing *religion?*

A. The Roman Catholic.

Q. What *minerals* are found here ?

A. Gold, silver, tin, and copper.

Q. Which are the principal *productions ?*

A. Cattle, and the precious metals.

West of Buenos Ayres is a plain 1500 miles *long, and* 500 *broad.* On this prairie innumerable herds of horses, mules, and cattle, *are found wild*, which are hunted for their hides and tallow.

Q. Which are the principal *exports ?*

A. The precious metals, hides, beef, and tallow.

Q. What is the *government ?*

A. In 1816, the country declared itself independent, and established a republican government.

Questions.—Map of S. America.

How are the United Provinces bounded N. E. S. W ? *Describe the great river of this country, and its branches.* What desert lies pa**''ly** in this country and where? Which is the chief town? Which the capital? *How are they situated? Where is Monte Video?* What two capes are at the mouth of the river La Plata?

Chili.

Q. How is Chili (Chee'le) *bounded?*

A. N. by the U... P...; E. by U... P... and P...; S. by P...; W. by the P... O...

1. The *face of the country* is mountainous, and the scenery picturesque and grand. The *soil* in the *south* is very rich, but in the *north*, dry and barren.

2. The *climate* is mild and healthful.

3. *Length*, 1300 miles; *breadth*, 140; *square miles*, 180,000. *Pop.* 1,200,000, exclusive of independent Indians.

Q. Which are the principal *mountains?*

A. The Andes, which separate Chili from the Provinces of La Plata.

Volcanoes are frequent in Chili. Not less than 14 *are in a constant state of eruption. Earthquakes* also occur several times in a year.

Q. Which are the principal *rivers?*

A. Chili has no large rivers, although it is extremely well watered by small streams.

Q. Which is the *chief town* and *capital?*

A. St. J...o, or S...o.

1. *St. J...o, 5th class, stands* on a beautiful plain. Its houses are of brick, and but of one story, on account of the earthquakes.

2. The *other principal towns* are Concep'tion, *6th class,* and Valparai'so, *large town,* 100 miles west of St. Jago, a very commercial city.

Q. What is the prevailing *religion?*

A. The Roman Catholic.

Q. What *islands* are on the coast of Chili?

A. Chiloe (Chil-lo-a') and Juan' Fernan'dez.

Chiloe is near the southern boundary of Chili. It is 18 miles *long. Juan Fernandez* is 300 miles west of Valparai'so.

Q. What *minerals* are found in Chili?

A. Gold, silver, and copper, particularly in the northern part.

Q. Which are the principal *productions?*

A. Wheat, wine, oil, and hemp.

Q. What remarkable *Indian nation* dwells in Chili ?

A. The Arauca'nians.

The *Araucanians are said to be* the most powerful and warlike of all the Indian nations, in the southern part of the continent, *resembling* the North American Indians in their character, particularly in their fondness for eloquence.

Q. What is the *government* ?

A. In 1818 it declared itself independent, and has established a republican government.

Questions.—Map of S. America.

How is Chili bounded N. E. S. W. ? What desert on the N. ? What islands on the W. ? What island near the southern part ? Which is the chief town and capital ? *How situated ? Where is Valparaiso ? Coquimbo ?*

Patagonia.

Q. How is Patagonia *bounded* ?

A. N. by c..., and the u... p...; E. by the a...; and S. by the St. of m...; W. by the p... and c...

1. The *Andes pass* through the western part, and render it mountainous. The eastern part is level.

2. The *climate* is cold, and the *soil*, so far as known, unproductive.

3. *Length*, 900 miles; *breadth*, 300; *square miles*, according to Hassel, 491,000. It is *inhabited* by Indians who are said to be very ferocious, and some of them of great stature. Their *number* is not known

Q. What *strait* lies south ?

A. The strait of Magel'lan.

Q. What *islands* lie near Patagonia ?

A. Terra del Fu'ego, Stat'en island, and Falk'land islands.

Q. On *what account* is this coast frequently visited ?

A. For seals, with which this coast, and the neighbouring islands, abound.

Questions.—Map of S. America.

How is Patagonia bounded N. E. S. W. ? What islands lie S of it ? What strait ? What islands E. ? What cape on the S. point of Terra del Fuego ? Where are the New South Shetland Islands ?

EASTERN CONTINENT.

Q How is the eastern continent *divided?*

A. Into Europe, Asia, and Africa, to which we may add Oceanica.

Questions.—*Map of the World.*

How is the eastern continent bounded N. E. S. W.? *Between what lines of latitude does it extend, reckoning from the Cape of Good Hope to Cape Taymour?* Which are its grand divisions?

EUROPE.

French.　　　　　*Spaniards.*

Dutch.　　　　　*Russians.*

Q. How is Europe *bounded?*

A. N. by the F... O...; E. by the A...; S. by the M... S... which separates it from Africa; W. by the A...

1. *Europe is smaller than either* Asia, Africa, or America.

2. The *south* of Europe enjoys a warm *climate;* in the *middle* it is temperate, and in the *northern regions* less severely cold, than in the same latitude in Asia and America.

3. The *length* of Europe, from the most western part of

Portugal, to the Uralian mountains, on the east, is 3300 miles; *breadth* from North Cape, in Lapland, to the southern extremity of Greece, is 2350 miles. *Square* miles, according to Hassel, about 3,250,000. *Pop.* from 180,000,000 to 200,000,000.

Q. Which are the principal *mountains?*

A. The Alps, the Pyr'enees, the Carpath'ian, the Ap'penines, Dofrafield, and U'ral mountains.

1. The *Alps* are the most celebrated mountains in Europe for their height and grandeur. *They divide* Italy from Germany, France, and Switzerland. *The length of the range* is 600 miles, and *the breadth*, in some places, exceeds 100. Among the *highest peaks* are Mont Blanc, and Mont Rosa, *3d class*, Mt. St. Bernard, *4th class*, and Mt. St. Gothard and Mount Cenis, *5th class*. The chain of Mount Jura *separates* Switzerland from France.

2. The *Pyr'enees separate* France from Spain, and *extend* from the Mediterranean to the bay of Biscay, a distance of 250 miles; their *breadth* varies from 50 to 100 miles Mt. Perdu, the *highest peak*, belongs to the *4th class*. The other *principal peaks* are Vignemale, Pic Blanc, and Pic Long.

3. The *Carpathian* mountains, *5th class, are in* Austria: they *separate* Hungary from Galicia; the *whole length* is 300 miles.

4. The *Ap'penines, 5th class, are a branch of* the Alps, which pass off at Gene'va, and traverse the whole length of Italy.

5. The *Dof'rafield, 5th class, separate* Sweden from Norway, and *extend* more than 1000 miles from north to south.

6. The *Ural* mountains, *5th class, are a part of the boundary between* Europe and Asia. They *extend* about 1500 miles.

Q. Which are the principal *seas?*

A. The Mediterra'nean, Baltic, North, Black, and White Seas, and the Archipelago, (Ar-ke-pel'-a-go.)

1. The *Mediterranean sea lies* between Europe, Asia, and Africa, and is the largest sea in the world, being 2000 miles *long* from east to west.

2. The *Baltic lies* between Sweden on the west, and Russia on the east.

3. The *North Sea lies* between Great Britain on the west, and Denmark on the east.

4. The *Black Sea lies* between Europe and Asia.

10*

'5. The *White Sea lies* in the northern part of Russia, and opens into the Frozen Ocean.

6. The *Archipelago lies* between Turkey in Europe, and Turkey in Asia.

Q. Which are the principal *channels?*

A. The English channel, St. George's channel, the Cat'tegat, and Skag'er Rack.

1. The *English channel lies* between England and France.
2. *St. George's channel lies* between England and Ireland.
3. *North channel lies* between Ireland and Scotland.
4. The *Cattegat* between Denmark and Sweden.
5. The *Skag'er Rack* between Denmark and Norway.

Q. Which are the principal *straits?*

A. The strait of Gibraltar, (Jib-rawl'ter) the Dardanelles, (Dar-da-nels') Strait of Constantino'ple, and Strait of Dover.

1. The strait of *Gibraltar connects* the Mediterranean and Atlantic.
2. The *Dardanelles connects* the Archipelago and the sea of Mar'mora.
3. The strait of *Constantinople connects* the sea of Marmora and the Black Sea.
4. The strait of *Dover connects* the North Sea and the English channel.

Q. Which are the principal *gulfs* or *bays?*

A. The Gulf of Ven'ice, Bay of Biscay, Gulfs of Both'nia, Finland, and Riga.

1. The *Gulf of Venice lies* between Turkey and Italy.
2. The *Bay of Biscay lies* north of Spain, and west of France.
3. The *Gulfs of Bothnia, Finland,* and *Riga,* are arms of the Baltic sea.

Q. Which are the principal *rivers?*

A. The Volga, Don, Dnieper, (Ne'per) Dniester, (Nees'ter) Danube, Rhine, Rhone, and Elbe.

1. The *Volga, 1st class,* is the largest river; it *rises* in Russia, and *flowing,* first easterly and then southerly, *empties* itself into the Caspian sea. It is *navigable* 2000 miles.
2. The *Don, 3d class, rises* in Russia, and *running* southerly, *flows* into the sea of Azof.
3. The *Dnieper, 2d class,* and the *Dniester, 3d class,* the *former* of which *rises* west of Moscow, and the *latter* in the Carpathian mountains; both *flow* southerly, and *empty* themselves into the Black sea.

·**4.** The *Danube, 2d class, rises* in Germany, through which *flowing* easterly, it *passes* into Hungary, where it *turns* south, then southeast, and *passing* into Turkey, it *empties* into the Black sea. It is navigable 1500 miles.

5. The *Rhine, 3d class, rises* in Switzerland, and after a circuitous *route, empties* itself into the North Sea.

6. The *Rhone, 4th class,* also *rises* in Switzerland, and *flowing* west, *falls* into the lake of Geneva. . Issuing from that lake, it pursues a south westerly *course* into France, where it *turns* south, and *flows* into the Mediterranean. It is the most *rapid* river in Europe.

7. The *Elbe, 3d class, rises* in Bohemia, and *flowing* in a north west direction, *enters* the North Sea.

Q. Which are the principal *capes ?*

A. North Cape, Naze, Land's End, Cape Clear, La Hogue, Finisterre, St. Vincents, and Mat'a-pan.

Q. Which are the principal *islands ?*

A. Sicily, Sardinia, and Corsica, in the Mediterranean ; Great Britain, Ireland, and Iceland, in the Atlantic ; Spitzber'gen and Nova Zembla, in the Frozen Ocean.

Q. What *countries* does Europe contain ?

A. Great Britain, France, Spain, Portugal, Italy, Tur'key, Switzerland, Neth'erlands, Ger'many, Aus'tria, Prussia, (Prus'hia) Russia, (Ru'shia) Po'land, Den'mark, Swe'den, Nor'way, and Lap'-land.

Europe contains 3 *empires,* arranged according to their population, as follows :—Russia, Austria, and Turkey ; 14 *kingdoms,* viz. France, Great Britain, Spain, Prussia, Naples, Netherlands, Sardin'ia, Sweden, Bava'ria, Portugal, Denmark, Wir'temberg, Han'over, and Sax'ony ; 3 *republics,* viz. Switzerland, Io'nian Islands, and St. Marino ; besides *several smaller states,* called Grand Duchies, and Principalities.

Questions.—*Map of Europe.*

How is Europe bounded N. E. S. W. ? Name particularly the seas on the S. and mountains and rivers on the E. ? What separates it from Africa and Asia ?—What mountains between Norway and Sweden ? What is their direction ? What mountains on the

northeast of Europe? Where are the Carpathian Mts.? Alps?
Appenines? Cevennes? Erzeberg? What mountains in Spain?
Where is Mt. Ætna? Vesuvius? Mt Hecla?—What seas are
in the N. of Europe? What are the eastern and northern gulfs of
the Baltic? What are the straits called that lead into the Baltic?
What five seas are in the S. of Europe? What four gulfs in the
Mediterranean? What strait leads into the Mediterranean? What
strait connects the sea of Marmora with the Black Sea? What the
sea of Marmora with the Archipelago? What the North Sea and
English Channel? What channel lies between England and
France? What between England and Ireland? What between Denmark and Sweden? What between Denmark and Norway? What
lakes are in Sweden? What two in Russia? Where is the White
Sea? What rivers flow into it? What river flows into the sea of
Azof? *What is its direction? Describe the four rivers which
empty into the Black Sea?* What river empties into the gulf of
Venice? What are the branches of the Danube? What rivers run
into the Baltic? Into the North Sea? Into the Bay of Biscay?
What rivers in Spain flow into the Mediterranean? What five
into the Atlantic?—What are the two largest islands W. of
Europe? What are the islands N. of Europe? Which are the three
largest islands in the Mediterranean? Which is the most easterly?
How is Sicily situated? Candia? Corsica? Where is Elba? Malta?
What islands are near the coast of Spain? Where is the Archipelago? What islands are in the Archipelago? What islands lie
northwest of the Morea? Which way is Ireland from Scotland?
What islands lie north, and what west of Scotland? What islands
are in the Baltic? What island south of Sicily?—What is the
most northern cape in Europe? The most southern? Where are
capes Finisterre, Land's End, Cape Clear, the Naze, St. Vincent,
La Hogue?—What countries does Europe contain?

Kingdoms of Great Britain and Ireland.

Q. What do the kingdoms of Great Britain and Ireland em-
brace?

A. England, Scotland, and Ireland, usually termed the British Isles.

1. *To Great Britain belong* numerous foreign possessions,
in Europe, Asia, Africa, America, and Oceanica, *containing*
more than 60,000,000 *inhabitants*. These will be noticed in
their proper place.

2. *Square miles*, 118,000. *Pop*. about 21,000,000.

Q. What is the *government* of the United Kingdoms?

A. A limited monarchy.

The supreme power is vested in a king and parliament.
The parliament consists of two houses, Lords and Commons;
the former are hereditary peers; *the latter* representatives,
chosen by the people.

ENGLAND AND WALES.

View of London

Q. How is England, including Wales, *bounded?*

A. N. by s...; E. by the N... s...; S. by the ST... of D... and B... C...; W. by the A..., ST... G... C..., and I... S...

1. *Julius Cæsar invaded Britain* 55 years before the Christian era. In the 5th century, *the Saxons conquered the country;* in the 8th the *Danes conquered* it; and in the 11th, *William of Normandy,* styled the conqueror. *The English are descended from* the ancient Britons, with a mixture of Saxons, Danes, and Normans. *The present royal family are descended from* William the Conqueror.

2. In some parts, the *face of the country* is rugged; but, *in general,* it is diversified with hills, and vales, and abounds with elegant scenery. *Wales* is mountainous.

3. The *soil* of England is, in general, excellent; and is under a higher *state of improvement* than any other country in Europe.

4. The *climate* of England is moist, and subject to frequent and sudden changes.

5. *Length,* 400; *average breadth,* 150; *square miles* of England and Wales, 58,000; *Pop.* 11,200,000; *Pop of Wales,* 717,000.

Q. Which are the principal *rivers?*

A. The Severn, Thames, (Temz',) Humber, and Mer'sey.

1. The *Severn, 5th class, rises* in North Wales, and *flowing easterly, south, and southeasterly, falls* into Bristol channel.

2. The *Thames, 6th class, rises* in the eastern part of the kingdom, and *flowing* southeast, *falls* into the North Sea.

3. The *Humber, 6th class,* is *formed* of the Ouse and the Trent; it *falls* into the North Sea.

4. The *Mersey, 6th class, flows* southwesterly, and *falls* into St. George's channel.

Q. What can you say of the *canals* in England?

A. They intersect the country in almost every direction.

1. More than 2400 *miles of artificial navigation have been formed,* in various parts of the kingdom. *One of the principal canals* is the *Grand Trunk,* which passes from the river Mersey 99 miles to the Trent, near the centre of the kingdom; thence 40 miles to the Severn. From the Grand Trunk, the *Oxford canal* extends 90 miles to that city. From the upper part of the Oxford canal, a branch of 100 miles is carried to the Thames, a short distance from London.

2. The *Ellesmere and Chester canal connects* the rivers Mersey, Dee, and Severn. It contains an aqueduct 1000 feet long, and 126 feet high.

Q. Which is the *chief town,* or city and *capital?*

A. L...n.

1. L...n, *1st class,* is the chief town and capital of the British empire. It is *situated* on the Thames, 60 miles from its mouth. Its *churches* are between 300 and 400; its *tonnage* more than half a million. It is principally *built of* brick. Its *public buildings are* the cathedral of St. Paul's, Westminster Abbey, and the Tower of London.

2. *York, 6th class,* is the *metropolis of* the North of England, and is second to London only in dignity.

3. The *commercial cities* are Liverpool, *3d class,* the next to London in commerce, and is distinguished for its trade with America; Bristol, *4th class,* the rival of Liverpool in commerce with America; Hull, *5th class,* the chief port on the eastern coast, north of London, distinguished for the whale fishery; and Newcastle, *5th class,* distinguished for its trade in coal.

4. The *principal manufacturing towns* are Manchester, *famous* for its cotton goods, and Birmingham for its toys, *3d class;* Sheffield for its cutlery, Leeds for its cloths, Norwich for its worsted stuffs, Leicester and Nottingham for their stockings, and hosiery, *5th class;* Coventry for its ribands, and Worcester for its woollens, and especially, for its porcelain, *6th class;* and Kid'derminster, *large town,* for its carpets.

5. The chief *naval stations* are Gosport, Portsmouth, and Chatham.

6. The *seats of literature* are Oxford and Cambridge.

7. The *places of fashionable resort*, for sea bathing, are Brighton, Ramsgate, and Margaté; *for the use of mineral waters*, Bath, Cheltenham, Clifton, Harrowgate, &c.

8. The *principal towns* in *Wales*, are Caermarthen, *large town*, in South Wales; and Caernarvon, *small town*, in North Wales.

Q. What *universities* does England contain ?

A. Those of Oxford and Cambridge.

Cambridge university contains 16 *colleges*, and more than 2000 *students*. It is *celebrated for* mathematical science. *Oxford* contains 25 *colleges*, and is *celebrated for* classical literature.

Q. What is the prevailing *religion ?*

A. Episcopacy is the established religion, but all others are tolerated.

Q. What is the *character* of the English ?

A. They are intelligent, frugal, brave, and industrious; but possess great national pride.

Q. Which are the principal *mines ?*

A. The tin mines of Cornwall; the coal mines of New Castle and its vicinity; and the rock salt mines, near Liverpool.

Q. Which are the principal *manufactures* of England ?

A. Woollen and cotton goods; articles of iron, tin, lead, and earthen ware.

Q. What can you say of its *commerce ?*

A. It exceeds that of any other country on the globe.

Q. What can you say of the British *navy ?*

A. It exceeds that of all the other nations of Europe, put together.

Q. What can you say of the *public debt ?*

A. In 1814, the public debt amounted to 700,000,000 of pounds sterling, and is increasing.

Q. What *islands* near the English coast belong to England ?

A. The Isle of Wight, on the southern coast; the small islands of Al'derney, Guern'sey, and Jer'sey, near the coast of France; the Isles of

Scil'ly, 30 miles west of Land's End; the Isle of Man, in the Irish Sea; and Angle'sea, on the coast of Wales.

Q. Which are the principal *productions?*

A. Wheat, barley, oats, and rye.

Q. Which are the principal articles of *export?*

A. Woollen goods, articles of iron, tin, lead, earthen ware, coal, and silk.

Questions.—Map of the British Isles.

How is England bounded N. E. S. W? What strait and channel lie between England and France? What channel and sea separate it from Ireland? What Island lies near the southern border of England? What near the north western border? What north of Anglesea? Where is Land's End? What islands lie west of Land's End? What islands belong to G. Britain near the coast of France? *Describe the five rivers, Severn, Thames, Humber, Mersey, and Tweed.* In what part of England is London? On what river? *What is its latitude? Where is Liverpool? In what direction from London? Which way from Liverpool is Manchester? Which way from Liverpool is York? Where is New Castle? Birmingham?* On what channel is b...tol? *In what direction from London? In what direction from Lon.. ...re Oxford and Cambridge? Where is Dover? Portsmouth? Plymou..* Which way from England is Wales? What channel lies between Wales and Ireland? Where is Cardigan Bay? *In what part c.. Wales is Caernarvon? Caermarthen?*

SCOTLAND.

Q. How is Scotland *bounded?*

A. N. by the A...; E. by the N... S...; S. by E... and the I... S...; W. by N... C... and the A...

1. Scotland was *united* with England in 1603. She is *represented in the British parliament* by 16 peers, and 45 members of the house of commons. *She, however, retains* her own ancient laws and judicial institutions.

2. The *northern part* of Scotland consists of barren hills and mountains, with numerous lakes. In the *southern part* it resembles England.

3. The *soil*, in general, is inferior to that of England, and is chiefly fitted for pasturage.

4. The *climate* is colder than that of England, but is more healthful.

5. *Length,* 275 miles; *breadth,* from 36 to 147; *square miles,* 30,000. *Pop.* 2;100,000.

Q. Which are the principal *mountains?*

A. The Grampian hills, which are the natural

boundary between the Highlands and Lowlands of Scotland.

Ben Nevis, 6th class, is the highest land in Great Britain.

Q. Which are the principal *rivers?*

A. The Tweed, which separates England from Scotland ; the Forth, Tay, and Clyde.

1. The Tweed, 6th *class, forms part of the boundary between* England and Scotland, and *falls* into the North Sea at Berwick.

2. The *Forth, Tay,* and *Clyde,* 6th *class,* all *fall* into Friths, of the same names.

Q. Which is the principal *lake?*

A. Loch, or lake Lomond, celebrated for its romantic scenery.

There are numerous other small lakes ; such as Loch Ness. Loch Oich, &c.

Q. Which is the *chief town* of Scotland ?

A. E...h.

1. E...h, 3d *class,* is about two miles from the Frith of Forth. The surrounding country is hilly, excepting towards the north. The New Town, as it is called, is built in the modern style, with great elegance. The houses of the Old Town are in some instances 14 stories high. The *sea-port* of Edinburgh is Leith, (Leeth) 6th *class,* two miles north of the town.

2. The *other principal towns* are Glas'gow, on the Clyde, distinguished for its commerce, manufactures, and literary institutions ; Pais'ley, St. An'drews, Dundee, 5th *class,* Perth, and Greenock, (Gren'ec) 6th *class,* which last is the principal sea-port of Scotland.

Q. What *universities* does Scotland contain ?

A. Those of St. Andrews, Aberdeen', Edinburgh and Glasgow.

1. The university of Edinburgh is the *most distinguished.* It has 27 professors, and 2000 students. *The medical school* attached to it is celebrated throughout Europe.

2. The *system of education* adopted in Scotland is celebrated throughout the world. All classes are well informed ; no people exceed them for morality and religion.

Q. What is the prevailing *religion?*

A. Presbyterian is the *established* religion.

Q. What is the *character* of the Scotch ?

A. They are hardy, enterprising, intelligent, virtuous, and good natured.

Q. Which are the principal *islands* belonging to Scotland ?

A. The Heb'rides, on the western coast; the Ork'neys on the north coast; and the Shet'land Islands, northeast of the Orkneys.

1 The *Heb'rides* are 300 in number, and contain 70,000 inhabitants.

2. The *Ork'neys* are 26 in number : Kirkwall, *small town,* is the *chief town.*

, 3. The *Shetland Islands* are 86 in number ; 40 of which are inhabited by 21,000 people. They own 70 or 80,000 sheep.

Q. Which are the principal *manufactures ?*

A. Cotton and linen goods.

Q. Which are the principal *productions ?*

A. Grass and oats; cattle and sheep are raised in vast numbers.

Questions.—Map of the British Isles.

How is Scotland bounded N. E. S. W. ? What river separates it from England ? *Describe the rivers Tweed, Forth, Tay, and Clyde.* In what part of Scotland are the Friths of Forth and Clyde ? Solway Frith ? Murray Frith ? Pentland Frith ? Where are the Hebrides ? What other islands on the western coast of Scotland ? What two clusters of islands N. of Scotland ? Which is the chief town, *and how situated ? Which way from Edinburgh is Glasgow ? What direction from Glasgow is Loch Lomond? Is Aberdeen ? What sea-port in the west of Scotland ?*

IRELAND.

Q. How is Ireland *bounded ?*

A. N. by the A...; E. by the N... C..., I... S..., and ST. G... C...; S. and W. by the A...

1. Ireland was *conquered* by England in the 12th century ; but was not *completely subdued,* till the 17th. The *legislative union took place* in 1807. Ireland *sends* 100 *representatives* to the house of commons, and 28 *peers* to the house of lords, besides five *spiritual lords.*

2. The *face of the country* is uneven, with hills of some height, but easy of ascent.

3. The *soil* is generally fertile ; the bogs and morasses, which cover one tenth of the surface, are unfit for cultivation.

4 The *climate* is temperate, cooler in summer, and warmer in winter, than that of England, but humid, and often foggy.

5. *Length,* 235 miles ; *greatest breadth,* 182 ; *square miles,* 30,000. *Pop.* 7,000,000.

Q. Which is the principal *river ?*

A. The Shan'non.

1. The *Shannon, rises* in the northern part of Ireland, *runs* southwest, and *enters* the Atlantic.

2. The *other principal rivers* are the Bar'row, which *rises* west of Dublin, and *flowing* southerly, *enters* Waterford harbour ; the Lif'fy, which *flows* into Dublin bay ; and the Boyne', which *rises* near the source of the Barrow, and *flowing* northeast, *empties* itself north of the Liffy, all of the 6*th class*.

Q. Which is the *chief town* ?

A. D...n.

1. *D...n*, 3*d class, stands* on the Liffy ; it is the second city in the united kingdom ; many of its edifices are magnificent. Its *harbour* is one of the most beautiful in Europe.

2. The *other principal towns* are Cork, 4*th class*, the second city in Ireland, distinguished for its foreign commerce ; Lim'erick, 4*th class*, Waterford and Belfast, 5*th class*.

Q. What *university* has Ireland ?

A. Dublin university, a celebrated institution.

It has a library of 70,000 volumes ; 13 professors, and 1500 students.

Q. Which is the prevailing *religion* ?

A. Episcopacy is the established religion ; but three quarters of the inhabitants are Roman Catholics, who, on account of their religion, are excluded from all offices under government.

Q. What is the *character* of the Irish ?

A. They are in general quick of apprehension, active, brave, and hospitable : but irascible, ignorant, and superstitious.

Q. What *curiosity* is on the northern coast ?

A. The Giants' Causeway.

The Giants' causeway *consists of* many hundred thousand columns of rock, of a dark gray colour, rising perpendicularly from 200 to 300 feet, from the water.

Q. Which is the chief article of *manufacture* and *export* ?

A. Linen.

Q. Which are the principal *productions*.

A. Potatoes, oats, and grass ; numerous herds of cattle are raised.

Questions.—*Map of the British Isles.*

How is Ireland bounded N. E. S. W. ? In what part of Ireland is Donegal Bay ? What island lies between Ireland and Scotland ?— What between Ireland and Wales ? Where is Cape Clear ? Which are some of the principal lakes ? *Describe the rivers Shannon*

Barrow, Liffy, and Boyne.   In what part of Ireland is the Giants Causeway? Which side of Ireland is Dublin? On what river?— *Where is Cork? Limerick? Galway? Londonderry?*

France.

View of Paris.

Q. How is France *bounded?*

A. N. by the B... C..., ST. of D... and N...; E. by N..., G..., S...; and I...; S. by the M... and S...; and W. by the B... of B...

1. *The French derive their name* from the Franks, *who conquered the country* in the 3d century.   In 1792, the *French revolution,* as it is called, *commenced;* in 1793, *Louis XVI. was condemned* and *executed,* and the regal government abolished.   In 1804, *Bonaparte was crowned emperor;* subsequently he greatly enlarged the limits of France.   In 1815, *he was overthrown* in the battle of Waterloo, and was *sent prisoner to* St. Helena, where, March 5th, 1821, *he died.* Louis XVIII. *was placed on the throne,* and the kingdom reduced to nearly its former limits.

2. The *face of the country* is generally level, or gently undulating.

3. The *soil* is generally fertile, but *agricultural improvements* fall much short of those of England.

4. The *climate* is temperate, dry, and salubrious.

5. *Length,* 650 miles; *breadth,* 500; *square miles,* 200,000. *Pop.* 30,000,000.

Q. Which are the principal *mountains?*

A. The Pyr'enees, Alps, Mount Jura chain, and Ceven'nes.

The *Pyrenees*, highest peaks, *4th class, separate* France from Spain. The *Alps*, highest peaks, 3d *class ;* other peaks *4th* and 5*th classes*, separate it from Italy. The Mount *Jura* chain, highest peaks, Reculet and Dole, 5*th class*, separates it from Switzerland. The *Cevennes*, highest peaks, 5*th class*, *run* parallel with the Rhone.

Q. Which are the principal *bays* or *gulfs?*

A. The Bay of Biscay, and the Gulf of Lyons.

Q. What *strait* deserves notice?

A. The strait of Dover, which divides France from England, and is 21 miles across.

Q. Which are the principal *rivers?*

A. The Rhine, Rhone, Garonne, (Garone') Loire, (Lwor) and Seine, (Sane.)

1. The *Rhine*, 3d *class, separates* France from Germany. The *Rhone*, 4th *class, rises* in Switzerland, *passes* through the lake of Geneva, and *flowing* southwest, and afterwards south, *empties* itself into the Mediterranean.

2. The *Garonne*, 4th *class, rises* in the Pyrenees, and *flowing*, upon the whole, in a northwest direction, *empties* itself into the Bay of Biscay. The *Loire rises* in the south of France, near the Cevennes, and *flows* north and west into the Atlantic.

Q. What *canal* can you mention?

A. The most celebrated, is the canal of Languedoc, which connects the Mediterranean, with the Bay of Biscay.

It is 140 miles *long*, 60 feet *wide*, and 6 feet *deep*.

Q. Which is the *chief town*, and *capital?*

A. P...s.

1. P...s, *1st class*, the second city in Europe, in size, is the first in splendour. It *lies* on the Seine, in the midst of an extensive and delightful plain. It *contains* many splendid palaces, and noble institutions. *Versailles*, (Ver-sails',) 6*th class*, in the *neighbourhood of Paris*, is a *considerable city*, and *contains a* palace which is a favourite residence of the Kings. *Havre de grace*, (*Haver de Gras*) 6*th class*, is the sea-port of Paris.

2. The *other principal towns* are Marseilles, (Mar-saiis') a noted sea-port on the Mediterranean ; and Lyons, famous for its silk manufactures, 3d *class ;* Bordeaux, (Boor-do',) the chief place for the export of wines, and Rouen (Roo'en) and Lisle, (Lele) manufacturing towns, 4th *class ;* Brest, the chief naval station, on the Mediterranean ; Rheimes,

(Rimes) celebrated in ancient times as the coronation place of the French Kings ; Amiens, (A'meens) remark.
able for an important treaty between England and France,
in 1802; and Strasburg and Orleans, manufacturing towns,
5th class.

Q. What *universities* does France contain ?

A. There are in France 26 universities, 36 royal colleges, besides 22,500 primary schools.

The *National institute*, a literary association at Paris, is the most learned scientific body in the world.

Q. What is the prevailing *religion?*

A. The Roman Catholic is the established religion ; but all others are tolerated.

Q. What is the *character* of the French ?

A. They are gay, lively, and polite, but inconstant and impetuous.

Q. What *Island* belongs to France ?

A. Cor'sica, in the Mediterranean, celebrated as the birth-place of Napoleon Bonaparte.

The *small islands* of Rha, (Ra) Belle-Isle, (Bel'ile) and Ush'ant, near the west coast, also *belong to France.*

Q. Which are the principal *manufactures?*

A. Silk, cloth, lace, and china.

Q. Which are the principal *productions?*

A. Wines, grain, and silk.

Q. Which are the principal *exports ?*

A. Wines, brandy, and manufactured articles of silk, gold, silver, and iron.

Q. What is the *government ?*

A. A limited monarchy, like that of Great Britain.

The *legislative power is vested* in the king, a house of peers, consisting of upwards of 200 members, and a house of delegates, consisting of 256 members, chosen by the peo ple.

Questions.—Map of Europe.

How is France bounded N. E. S. W. ? What mountains separate
it from Spain ? What from Italy ? What chain runs parallel with
the Rhone ? What two rivers empty into the Bay of Biscay ?—
What river empties into the British Channel ? What into the
Mediterranean ? *Where do these rivers rise? What is their
course ?*—Which is the chief town and capital ? *On what river does*

it stand? Where is Lyons? Marseilles? Bordeaux? Brest?
Calais?—What is the latitude of Paris? Which way from Paris
is London? Which way from Paris is Madrid? Petersburg?
Where is the I. of Corsica which belongs to France?

Spain.

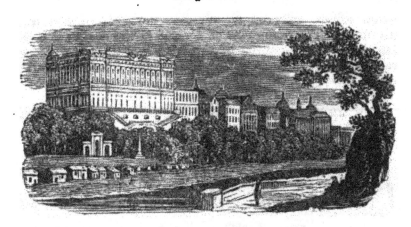

View of Madrid.

Q. How is Spain *bounded?*

A. N. by the B... of B... and F...; E. by the M...
S...; S. by the M... S..., ST. of G... and the A...;
W. by P... and the A...

1. *Spain was conquered* by the Romans under Augustus
Cæsar. In the 8th century the *Moors took possession* of the
country, but *were expelled* in the 17th century. In the 16th
century *Spain was esteemed the most formidable* power in
Europe. *Since that time,* her power has declined, and *at
present* she is in an unsettled *state.*

2. The *face of the country* is generally diversified by mountains and valleys, and adorned with fine scenery.

3. The *soil* is generally fertile, but the state of cultivation
is poor.

4. The *climate* of the south of Spain is very hot; but the
air is dry and serene; *the interior* on account of its elevation,
is cool.

5. *Length,* 650 miles; *breadth,* 550; *square miles,* 176,000;
Pop. 11,200,000.

Q. Which are the *principal mountains?*

A. The Pyr'enees, which separate Spain from
France, together with their branches.

·*These branches* are the Cantab'rian chain which *runs west,* parallel with the northern coast, and *terminates* at cape Finis-terre. The *Ibe'rian range*, highest peaks, 5th c*lass*, *springs* from the middle of the Cantabrian chain, and *runs* first S. E. and afterwards S. till it reaches the Mediterranean.— The *mountains of Castile*, (Cas-teel') the *mountains of To-le'do*, and the *Sierra Morena*, 6th class, *all spring from* the Iberian range, and *run* S. W. into Portugal parallel with each other. The mountains of Castile in the *north ;* of Tole-do in the *middle ;* and the Morena in the *south*. The *Sierra Niva'da*, highest peaks, 4th class, *springs from* the Iberian range, and *runs* S. W. to the strait of Gibraltar.

Q. Which are the principal *capes ?*

A. Capes Ortegal' and Finisterre', in the north-west ; and cape Trafal'gar, in the southwest.

Q. What *strait* is worthy of notice ?

A. The strait of Gibraltar, (Jib-rawl'tar) which separates Spain from Africa.

1. This strait is 15 miles *broad ;* it forms the entrance into the Mediterranean sea.

2. The celebrated *Fortress of Gibraltar* is possessed by the British. It is a mere rock, 1400 feet high ; but, from its peculiar structure, has been rendered almost an impregna-ble fortress. The *town is situated* on the declivity of the rock, and is a place of great trade.

Q. What *islands* belong to Spain ?

A. Major'ca, Iv'ica, and Minor'ca.

1. *Majorca* is the largest of these islands. It is fruitful, and is surrounded with watch towers. Its *capital* is of the same name.

2. *Minor'ca* has an excellent harbour—Port Mahon'.—The island *yields* vines, olives, and cotton. Citadella is the *capital*.

3. *Iv'ica produces* corn, wine, and other fruits. It is noted for the great quantity of salt made in it. Its *capital* is Ivica.

Q. Which is the principal *river ?*

A. The Ta'gus.

1. The *Tagus*, 4th class, *rises* in the Iberian mountains, and is *navigable* 50 or 60 miles. It *passes* westerly through Portugal into the Atlantic.

2. The *Guadiana*, 4th class, *rises* in some lakes in New Castile, and *flowing* southwesterly, *enters* the Atlantic be-tween Spain and Portugal ; the *Guadalquiver*, (Gwa-dal kee'ver) 4th class, *rises* in the Sierra Nivada mountains, and *flowing* S. W. *enters* the Atlantic N. W. of Cadiz ; the *Ebro*, (A'bro) 4th class, *rises* in the Iberian mountains

and *running* S. E. *empties* into the Mediterranean; the *Douro, 4th class, rises* in the Iberian mountains, and *flowing* westerly through Portugal, *empties* itself into the Atlantic, below Oporto; and the *Minho, 6th class,* which *rises* in the W. part of the Cantabrian chain, and *flowing* southwesterly, *enters* the Atlantic.

Q. Which is the *chief town* and *capital?*

A. M...d.

1. *M...d, 2d class, is situated* in a steril region, in the centre of Spain, on a branch of the Tagus, 2000 feet above the level of the sea. It is a place of but little trade, but derives its subsistence chiefly from remote provinces, or foreign countries. The royal palace is magnificent.

2. The *other principal towns* are Cadiz, *4th class,* in the southwest, the first commercial city of Spain; Barcelo'na, *3d class,* on the Mediterranean, the second commercial city; also on the same coast. Valencia, *4th class, celebrated* for its commerce and manufactures; and Malaga, *4th class,* for its wines.

3. The *principal towns in the interior* are Seville, (Saveel'ya) *4th class,* on the Guadalquiver, once the first town of Spain; Granada, *4th class,* east of Seville; Saragos'sa, *4th class,* famous for the siege by the French in 1809; To-lo'do, *5th class,* south of Madrid; Salaman'ca, *6th class,* N. W. of Madrid.

Q. What *Universities* can you mention?

A. Those of Salamanca and Saragossa, are the principal.

The university of Salamanca is the most *celebrated.* It has 61 professors, and formerly had 15,000 students. Yet, *education in Spain* is at a very low ebb, and the instruction imparted, is mingled with many superstitious and antiquated notions.

Q. What is the prevailing *religion?*

A. The Roman Catholic is the established religion; no other is tolerated.

Q. What is the *character* of the Spaniards?

A. They are grave, temperate, and polite; indolent, proud, bigoted, and very revengeful.

Q. What is the favourite *amusement* of the Spaniards?

A. The bull fights, or bull baiting.

These fights take place in amphitheatres, prepared for the purpose. The animal is first attacked by horsemen, armed with lances; then by men on foot, who carry a kind of ar-row like a fish hook, which gives the animal great pain, and redoubles his fury. When the bull is almost exhausted, a man called the *matador* advances, with a long knife, and generally, kills him with a single blow.

Q. Which is the principal *manufacture?*

A. Silk.

Q. Which are the principal *productions?*

A. The olive, the vine, figs, lemons, and grain in abundance ; wool is the great staple of the northern and central provinces.

Q. Which are the principal *exports?*

A. Silk, wool, wine, and raisins.

Q. What is the *government?*

A. At present, it is a limited monarchy.

Questions.—Map of Europe.

How is Spain bounded N. E. S. W. ? What chains of mountains can you mention ? What chain separates it from France ? What river has it on the E. ? What river empties S. of Cadiz ? What two rivers partly form the boundary between Spain and Portugal ? What two rivers run westerly, and pass into Portugal ? What two capes are on the northern coast ? Which is the capital, *and how situated? In what latitude is it? Where is Cadiz? Malaga? Carthagena? Valencia? Where is Seville, the ancient capital? Which way from Madrid is Cadiz? Where is the British fortress of Gibraltar?*

Portugal.

View of Lisbon.

Q. How is Portugal *bounded?*

A. N. and E. by s... ; S. and W. by the ▲...

1. The *surface* of Portugal is for the most part agreeably diversified; some portions, however, are rugged and stony.

2. The *soil* is fertile, especially in the valleys.

3. The *climate* is pleasant and healthful.

4. *Length*, 350 miles; *breadth*, 120; *square miles*, 35,000 *Pop.* 3,200,000.

Q. Which are the principal *mountains?*

A. The mountains of Castile, of Toledo, and of Sierra Morena, which come from Spain, and pass through Portugal, to the Atlantic.

Q. Which are the principal *rivers?*

A. The Min'ho, Dou'ro, Ta'gus, and Guadia'-na, all of which come from Spain, and *flow* westerly into the Atlantic.

Q. What *cape* is worthy of notice?

A. Cape St. Vincent.

Q. Which is the *chief town* and *capital?*

A. L...n.

1. *L...n, 2d class*, is at the mouth of the Tagus; it has a fine harbour, and an extensive commerce. Lisbon wine comes from this city.

2. The *other principal towns*, are Oport'o, *4th class, situated* on the Douro, famous for its port (or Oporto) wine; St. Ubes, which has a considerable trade in salt; and Coim'-bra, which is celebrated for its university, both *6th class*.

Q. What *university* does Portugal contain?

A. The university of Coimbra, which has 800 students, and many professors.

Q. What is the prevailing *religion?*

A. The Roman Catholic.

Q. What is the *character* of the Portuguese?

A. They are said to be friendly and hospitable; but are superstitious, haughty, and indolent.

Q. Which are the principal *productions* and *exports?*

A. Wool, wine, silk, and fruit.

Q. What is the *government?*

A. An absolute monarchy.

Questions.—Map of Europe.

How is Portugal bounded N. E. S. W.? What two rivers partly form the boundary between Portugal and Spain? What rivers from

Spain pass through Portugal, *and near what towns do they empty?* What cape is in the southwestern part of Portugal? What is the chief town and capital? *In what direction from Lisbon is Madrid? Where is Oporto?*

Italy.

View of Rome.

Q. How is It'aly *bounded?*

A. N. by s..., a..., and g... of v...; E. by the g... of v...; S. by the m...; W. by the m... and f...

1. The *face of the country* is diversified, and is much celebrated for the beauties of its landscape.

2. The *soil*, in general, is very fertile, and under high cultivation.

3. The *climate* is temperate, and salubrious.

4. *Length*, about 700 miles; *breadth*, very unequal—on the *north*, 350; in the *middle*, 140; and at the *southern extremity*, 75, *square miles*, 117,000. *Pop.* 19,245,000.

Q. Which are the principal *mountains?*

A. The Ap'penines, which traverse nearly the whole length of Italy.

Q. What remarkable *volcanic mountains* can you mention?

A. Mount Ætna, in Sicily, and Vesuvius, near Naples.

1. *Ætna, 4th class,* is about 180 miles in *circumference.* Its lava sometimes *flows* to the distance of 30 miles. Large stones have, at times, been thrown to the height of 7000 feet.

2. *Vesuvius* is of the *6th class,* but its explosions are tremendous. The *first eruption on record,* is that of the year

79, when the two cities Pompeii and Herculaneum were buried in lava and ashes. Vesuvius is 30 miles in *circumference*, and its crater, or aperture, about half a mile.

Q. Which is the principal *river?*

A. The Po.

1. The *Po, 3d class, rises* in the Alps, on the borders of France, and *flowing* east, *enters* the Gulf of Venice.

2. The *other rivers*, are the Adige, (A-dizh') 5*th class*, and Bren'ta, 6*th class*, both of which *fall* into the Gulf of Venice, and the Ar'no, and Ti'ber, 6*th class*, which *enter* the Mediterranean.

Q. Which are the principal *gulfs?*

A. Those of Venice, (Ven'is) Taren'to, Gen'oa, and Na'ples.

Q. Which are the principal *straits?*

A. The strait of Messi'na, between Italy and Sicily; and Bonifa'cio, (Bo-ne-fah'che-o) between Cor'sica and Sardin'ia.

Q. Which are the principal *universities?*

A. Those of Padu'a, Pavia, and Pisa.

Q. What is the prevailing *religion?*

A. The Roman Catholic.

Q. What is the *character* of the Italians?

A. Very different from their ancestors, the Romans, who conquered and governed the world; the modern Italians are effeminate, superstitious, bigoted, and slavish.

Q. What *islands* lie near Italy?

A. Sic'ily, Sardin'ia, Cor'sica, El'ba, Malta, (Mawl'ta.)

Corsica belongs to France; *Malta* has a *population* of 70,000 inhabitants, and *belongs* to Great Britain.

Q. Which are the principal *productions* and *exports?*

A. Fruits, wine, corn, silk, oil, and fine marble.

Q. How is Italy *divided?*

A. Into 9 states, viz. Lom'bardy, or Austrian Italy; kingdom of Sardinia; Duchies of Modena (Mo-da'na) of Luc'ca; of Par'ma; Grand Duchy of Tuscany; Roman states, or Pope's Dominions; Republic of St. Marino, (Ma-ree'no) and the kingdom of the two Sicilies.

Q. What is the *government* of these states ?

A. They are absolute monarchies, except the Republic of St. Marino.

1. LOMBARDY, OR AUSTRIAN ITALY.

Q. How is Lombardy, or Austrian Italy, *situated?*

A. It lies between the River Po, on the south, and the Alps, on the north.

1. *Lombardy* is a part of the Austrian dominions.
2. It is *divided* into the Territories of Venice and Mantua.
3. *Square miles*, 18,000. *Pop.* about 4,000,000.

Q. Which is the *chief town?*

A. V...e.

1. *V...e, 5th class*, is built on 72 islands, at the head of the Gulf of Venice. It has greatly declined in its commercial greatness, its manufactures, and population.
2. The *other principal towns*, are Mi'lan, *3d class*, distinguished for its manufactures and trade ; Verona and Padua, *5th class*, and Mantua, *6th class*, distinguished as the birth place of Virgil.

2. KINGDOM OF SARDINIA.

Q. How is the kingdom of Sardin'ia *situated ?*

A. The continental part of it lies on the north-western part of Italy, and is separated from France by Switzerland and the Alps. This, and the Island of Sardinia, make up the kingdom.

1. The kingdom of Sardinia is *divided* into Savoy', Piedmont, Gen'oa, and the Island of Sardinia.
2. *Square miles*, 27,000. *Pop.* about 4,000,000.

Q. Which is the *chief town* and *capital ?*

A. T...n, in Piedmont, (Pee-a-mont'.)

1. *T...n, 4th class*, is a strongly fortified city on the Po. It is *distinguished* for its manufactures of silk.
2. The *other cities* are Gen'oa, *4th class*, the birth place of Columbus, at the head of the gulf of Genoa. Its *commerce* is very extensive. Cagliari, (Cal'ya-ry) *5th class*, is the principal town of the Island of Sardinia.

3. MODENA. 4. LUCCA. 5. PARMA.

Q. How are the Duchies of Modena, Lucca, and Parma *situated?*

A. Nearly in the middle of Italy.

Q. Under whose *government* are these Duchies ?

A. The Duchy of Modena is under the government of the Arch Duke Francis, of the House of Austria; the others are under the government of Maria Louisa, wife of Napoleon Bonaparte, after whose death they revert to Spain.

Q. Which are the *chief towns* of these Duchies?

A. Modena, (Mo-da'na) Par'ma, and Lucca.

1. *Modena*, *6th class*, is 30 miles south of Mantua; it has a university with a fine library.

2. *Parma*, *5th class*, is west of Modena on the Po; *Lucca*, *6th class*, is N. E. of Pisa.

6. GRAND DUCHY OF TUSCANY.

Q How is Tuscany *situated?*

A. On the Mediterranean, northeast of the Pope' dominions.

1. *Tuscany* is the most prosperous of the Italian states.

2. The *territory* is small, but *fertile*, and well cultivated.

3. *Square miles*, 8500. *Pop.* 1,180,000.

Q. Which is the *chief town* and *capital?*

A. F...e.

1. *F...e*, *4th class*, *lies* on the Arno, and is one of the principal cities in Italy, and one of the handsomest in Europe. It abounds with elegant statues and paintings.

2. The *other principal towns*, are Leghorn', *4th class*, on the coast south of the Arno, *famous* for its straw hats; and Pisa, *6th class*, an ancient but decayed town.

3. The little island of *Elba*, which belongs to this Duchy, is fertile and salubrious. It will *long be known* as the temporary residence of Napoleon Bonaparte.

7. ROMAN STATES, OR POPE'S DOMINIONS.

Q. How are the Roman states *situated?*

A. They lie between Naples on the south, and the kingdoms of Tuscany and Austrian Italy, on the north.

1. The *Roman states* are Rome and Bologna, (Bo¹lone'ya.)

2. The *soil* is fertile, and *climate* warm.

3. *Square miles*, 14,500; *Pop.* 2,425,000.

Q. Which is the *chief town* and *capital?*

A. R...e.

1. *R...e*, *3d class*, was once a magnificent city, and the mistress of the world. The Pope still resides here, and besides his temporal jurisdiction over the Roman states, claims authority over the Catholic church throughout the world;

his influence, however, is now small. The city lies on
the Tiber, 15 miles from its mouth. The remains of columns,
temples, amphitheatres, &c. bespeak its former magnificence.
The church of St. Peter's is the largest in the world; it is
730 feet long, 530 broad, and 450 high; it was finished in
1620, having been upwards of 100 years in building. The
Vatican is a palace, containing 4000 apartments, and has a
library of more than 500,000 volumes. The population of
Rome was once near 7,000,000.

2. The *other principal city*, is Bologna, *4th class*, the seat
of a renowned university, and a celebrated academy of science.

8. REPUBLIC OF ST. MARINO.

Q. How is the republic of Marino (Ma-ree'no) *situated?*

A. It occupies a mountain and a surrounding district of 40 miles in the Roman territory.

1. The inhabitants *elect their magistrates*, but are under
the Pope's protection.
2. The *population* is 7000. The *inhabitants* are a simple,
industrious people, who have maintained their independance
for 1000 years.

9. NAPLES, OR THE TWO SICILIES.

Q. How is the kingdom of the two Sicilies *situated?*

A. It lies south of the Pope's dominions, and includes the island of Sicily.

1. The *face of the country* is variegated and mountainous.
2. The *soil* is very fertile, and the *climate* warm.
3. *Square miles*, 42,000.   *Pop.* 6,673,000.

Q. What is the *chief town* and *capital?*

A. N...s.

1. *N....s, 2d class*, is the fourth city in Europe, in point
of population, and surpasses every other in its situation and
appearance; but it is not remarkable for its commerce, or
manufactures.
2. Palermo, *3d class*, is the *chief town of the island of Si
cily*. It carries on an extensive commerce. Its silk manu
factures, it is said, employ 900 looms.

Questions.—Map of Europe.

How is Italy bounded N. E. S. W.? What river empties into the
G. of Venice? *What is its direction?* What mountains run through
Italy? Where is the kingdom of Naples? What is the capital? What
volcano near this city? Where is Sicily? What strait separates it
from Italy? What volcano is on this island? Where are the Lipari
islands? *Which way from Naples is Rome? Where is Florence?*

Leghorn? Genoa? What G. lies on the E. of Italy? What S. E.? Which way from Italy are Corsica and Sardinia? What strait separates these islands? What island lies between Corsica and Italy? In what direction from Rome is London? What is the latitude and longitude of Rome? What is the difference between the latitude of Rome and Washington?

𝕿𝖚𝖗𝖐𝖊𝖞 𝖎𝖓 𝕰𝖚𝖗𝖔𝖕𝖊.

View of Constantinople.

Q. How is Turkey in Europe *bounded?*

A. N. by the A... D..., and R...; E. by R..., the B... S..., S... of M..., and the A...; S. by A... and the M...; W. by the M..., G... of V... and A...

1. The *face of the country* is mountainous, interspersed with beautiful and fertile valleys.

2. The *soil* is very fertile.

3. The *climate* is generally mild, regular, and healthful.

4. *Length,* 870 miles; *square miles,* 206,000; *Pop.* 10,000,000, *consisting principally* of Greeks and Turks.

Q. Which are the principal *mountains?*

A. The Carpath'ian, between Turkey and Hungary, Mount He'mus, Olym'pus, Pe'lion, Parnas'sus, and Hel'icon.

Q. Which are the principal *seas?*

A. The Black Sea, Sea of Mar'mora and Archipe'lago, (Ar-ke-pel'a-go) which separate Turkey in Europe from Turkey in Asia.

Q. Which are the principal *straits?*

A. The straits of Dardanelles' and of Constantino'ple. 12*

The *Strait of Constantinople* is between the Black Sea and Sea of Mar'mora; the *Dardanelles* is between the Sea of Mar'mora and the Archipe'lago.

Q. What do you understand by the *Morea?*

A. It is a large peninsula in the south of Turkey, connected to the main land, by the Isthmus of Cor'inth, and inhabited principally by Greeks.

Q. Which is the principal *river?*

A. The Dan'ube.

1. The *Danube,* 2d *class, rises* in the southwest corner of Germany, whence, *running* easterly through Germany, it passes into Hungary, where it *turns south, and then south* east, and *passing* into Turkey, *flows* into the Black Sea.

2. The *Save,* 4th *class,* is a *branch* of the Danube; it *rises* in Germany and *flows* easterly; the *Pruth,* 5th *class, rises* in Galicia, and *flowing* southeasterly *falls* into the Danube. The *Narizza,* 5th *class, rises* near Mount Hemus, and *flows* southerly into the Archipelago. The *Var'dar,* 5th *class, rises* in Macedonia, and *running* southeast *falls* into the Gulf of Saloni'ca.

Q. Which is the *chief town* and *capital?*

A. C...e.

1. C...e, 1st *class, lies* on the strait of Constantinople, sometimes called the Bos'phorus; it is 25 miles in circumference.

2. The *other principal towns,* are Adriano'ple, 3d *class,* on the Marizza, a place of considerable commerce. Sophia, 4th *class,* which is next to Adrianople; Bel'grade, 6th *class,* a frontier port, noticed for its fortifications; and Saloni'ca, 4th *class,* next to Constantinople, in commercial importance.

Q. What is the prevailing *religion?*

A. The Turks are Mahomedan; the Greeks are Christians, under the Patriarch of Constantinople.

Q. What is the *character* of the Turks?

A. They are ignorant, haughty, indolent, and intolerant; but honest, and hospitable to strangers.

A striking mark of Turkish hospitality is seen in the caravan'saries, or public inns, which are to be met with in almost every village. In these, travellers *may remain* 3 days, *without expense.*

Q. What is the *character* of the Greeks?

A. They are active, lively, and courteous, but avaricious, treacherous, and insincere.

Q. What is the *government* of Turkey?

A. An absolute monarchy.

1. The Turkish *empire* is sometimes *called* the Ottoman empire, from Ot'toman, or Oth'oman, a prince of the Turks, who laid the foundation of the empire, 1299, A. C., and assumed the title of Sul'tan.

2. The *emperor is called* Sultan, or Grand Seignor; his *principal minister is called* Grand Vizier; the *governors of the province are called* Pa'chas, or Bashaws'. The *Turkish soldiers are called* Jan'isaries.

3. *Ancient Greece occupies* the southern part of Turkey in Europe; it was at length *conquered* by the Turks, who, for centuries, have exercised a most despotic sway over the subjugated Greeks. In 1820, the Greeks of the More'a, and adjacent islands and provinces, *threw off the Turkish yoke,* and are (1826) struggling for their independence.

Q. Which are the principal *manufactures?*

A. Carpets, muslins, crapes, and cannon.

Q. Which are the principal *productions?*

A. Corn, wine, oil, figs, and wool.

Q. Which are the principal *exports?*

A. Both articles of manufacture, and the productions of the soil.

Q. Which are the principal *islands* of the Archipelago?

A. Can'dia, Ne'gropont, and Lem'nos.

1. *Candia* is the *largest;* it was *anciently called* Crete'. It has a fertile *soil* and fine *climate.* Ne'gropont is the *next in size;* the *interior* is mountainous; the *soil* fertile. Lem'-nos is about 15 miles *long,* and 11 *broad. Pop.* about 8000.

2. On the *European* side are several islands, *called* Cyc'-lades. The *largest* is An'dros; *next to this are* Nax'os—Paros, *whence comes* the Parian marble; Antipa'ros, *celebrated* for its grotto, and Santori'ni.

3. On the *Asiatic* side are Mityle'ne, the ancient Les'bos, and Ten'edos, well known in Trojan history. Sci'o is 100 miles in *circumference.* In 1822 it contained 100,000 Greeks, and was flourishing; but the Turks *have desolated it,* and *massacred the inhabitants.* Pat'mos is a barren rock, on which John wrote the Book of Revelation. Rhodes' was once famous for its commerce and naval power. Cy'prus, on the south coast of Asia Minor, is about 150 miles *long,* and 70 *broad.* It was once distinguished for its fertility and population. It is now neglected and nearly depopulated.

Questions.—Map of Europe.

How is Turkey bounded N. E. S. W. ? Which are the principal rivers ? *Where do they rise? What is their course and place of discharge?* What mountains can you mention ? What is the capital ? *Where situated? Which way from Constantinople is Adrianople? Where is Sophia? Saloni'ca?* What sea lies soutn of Turkey ? What country lies west of this sea ? What is the name of the peninsula south of Greece ? What sea and strait between Constantinople and the Archipelago ? What large Island south of the Archipelago ? What cape at the southern extremity of the Morea ? What islands does the Archipelago contain ?

Ionian Republic.

Q. What do you understand by the *Ionian Republic?*

A. It consists of seven islands, on the W. coast of Greece, under the protection of Great Britain.

1. *These islands are* Cephalonia, the largest, Cor'fu, the *seat of government,* Zante, Cerigo, Ith'ica, Santa Mauro, and Paxo.

2. The *inhabitants* are Greeks and Italians, who are represented as intelligent and enterprising, and who carry on a considerable commerce in wine, oil, and spirits, the produce of the soil.

3. They are *governed* by a president, or commissioner, appointed by the king of England.

Switzerland.

View of Berne.

Q. How is Switzerland *bounded?*

A. N. by F... and G...; E. by A...; S. by I...; and W. by F...

1. The *face of the country* is mountainous, with deep valleys, and beautiful lakes, rendering the scenery often wild, and picturesque.

2. The *soil* in the valleys is fertile, but much of the country is uninhabitable.

3. The *climate* of Switzerland, varies in different parts. On *the mountains*, it is cold, while the heat in *the valleys* is often excessive. *Changes in the weather* consequently are frequent and great.

Length, 200 miles; *breadth*, 140; *square miles*, 18,000; *Pop.* 1,700,000.

Q. Which are the principal *mountains?*

A. The Alps.

1. The most *elevated parts* are Mount Blanc, *3d class*, or White Mountain, so called from the immense mantle of snow, of glaring whiteness, with which its summit and sides are always covered; and St. Goth'ard, *5th class*.

2. In the elevated hollows of the Alps, or between the peaks, are to be found what are called *glaciers; they consist of* vast masses of ice, frequently from 16 to 20 miles in length, and 100 to 600 feet deep; presenting to the eye an extensive mirror of ice, in some cases smooth, and unbroken, and in others, crowned with frightful chasms, and are adorned with pinnacles of ice, rising in various forms, and appearing like the spires and turrets of a city of crystal.

3. *Avalanches are* immense masses of snow and ice, which occasionally detach themselves from these glaciers, and are precipitated down the Alps, into the valleys, with a tremendous roar; sometimes levelling forests, and overwhelming villages, in their desolating course.

Q. Which are the principal *lakes?*

A. The lakes of Con'stance, Gene'va, Zurich, (Zu'rick) and Neufchatel, (Noo-sha-tel'.)

Q. Which are the principal *rivers?*

A. The Rhine and Rhone.

1. The *Rhine*, *3d class*, rises in Mt. St. Gothard, and *running* northeasterly *falls* into lake Constance; thence *running* westerly, and again northerly, *separates* France from Germany, and *passing through* the Prussian *dominions* and the Netherlands, *flows* westerly into the North Sea.

2. The *Rhone*, *4th class*, *rises* near Mt. St. Gothard, and *passing* westerly through the lake of Geneva, *turns* southwesterly into France, through which it *flows* southerly into the Mediterranean.

Q. Whit a is the *chief town*, and which the *capital?*

A. G...a is the chief town; B...e is usually considered the capital.

1. *G...a, 6th class,* is beautifully *situated* on the lake of Geneva; it is *celebrated for* having been the residence of Calvin, the reformer.

2. *B...e, 6th class,* is on the Aar.

3. The *other principal towns* are Ba'sil and Zu'rich, *6th class.*

Q. Which is the principal *university?*

A. The University of Geneva.

1. This university has 22 *professors,* 1000 *students,* and a *library* of 50,000 volumes.

2. There is also a *celebrated university* at Zurich, and *colleges* at Berne, and Basil.

Q. Which are the principal *religious denominations?*

A. Calvinists and Roman Catholics; the Calvinists are the most numerous.

Q. What is the *character* of the Swiss?

A. They are, simple, industrious, brave, independent, and strongly attached to their native soil.

Q. Which are the principal *productions?*

A. They cultivate some grain, but insufficient for consumption. The raising of cattle is their chief employment.

Q. What is the *government* of Switzerland?

A. The country is divided into 19 cantons, each of which is an independent republic, but united for the common safety, and are governed by a Grand Diet, or Confederative Assembly.

Questions.—Map of Europe.

How is Switzerland bounded N. E. S. W.? What two lakes do you notice? What two large rivers rise in S. and pass through those lakes? Which is the capital, *and how situated? In what direction from Berne is Geneva? Near what lake is Geneva situated?* What mountains separate S. from Italy?

Netherlands.

View of Amsterdam.

Q. How are the Netherlands *bounded?*

A. N. by the N... s...; E. by G...; S. by F...; W. by F. . and N... s...

1. The *kingdom of the Netherlands includes* the country formerly called Holland, or the Seven United Provinces. *Holland was formerly* a republic; *the Belgic provinces,* or Netherlands, *have been possessed* by various powers. They were *united* in 1814.

2. This is the most *level country* in Europe; it *lies* lower than the level of the sea, at high water mark, and is *protected from being overflowed* by high mounds, or dikes.

3. The *climate* on the coast is damp, and rather unhealthful; but in the *interior* is more dry, and agreeable.

4. The *soil* is in general fertile.

5. *Square miles,* 24,000. *Pop.* 5,200,000.

Q. Which are the principal *rivers?*

A. The Rhine, Scheldt, (Shelt) and Meuse.

1. The *Rhine, 2d class, comes from* Germany, and divides into several branches; *one branch flows* into the Zuy'der Zee; the *rest* into the North Sea.

2. The *Scheldt* and *Meuse* both *rise* in France, and *run* northerly. The *former, 5th class, flows* into the North Sea, and the *latter, 4th class, joins* the Rhine.

Q. Which is the *chief town,* and which are the *capitals?*

A. A...m is the chief town; B...s and the H...e are the two residences of the king, and his court

1. *A...m, 2d class*, is the *chief town* of North Holland, and the largest, richest, and most populous city of the Netherlands. It is on an arm of the Zuyder Zee. It is built on upwards of 13,000 piles. It *contains* many splendid edifices, and was formerly the second commercial city of Europe.

2. *B...s, 3d class*, and the *H...e, 4th class*, in South Holland, are among the most elegant towns in Europe. Brus'sels is *famous* for its lace, camlets, and carpets.

3. The country contains many other large towns. Among which are Ant'werp, formerly a great commercial city; Rot'-terdam, famous as the birth place of Erasmus; Ghent, (Gent) Leige, (Leege) and Bruges, all *4th class*, and Tournay, ('Toor-nay') noted for trade and manufactures; Haerlem, ('Har-lem') *6th class*, for its trade in flowers, and for its organ; Dort, *6th class*, for its synod; Schiedam, *large town*, for gin; and Spa, *small town*, for mineral waters. At Waterloo, *small town*, nine miles south of Brussels, was fought, 18th of June, 1815, the famous *battle*, which decided the fate of France and of Europe, and which led to the downfall of Bonaparte.

Q. What is the prevailing *religion?*

A. The inhabitants of the Provinces of Holland are mostly Calvinists; those of the Netherlands chiefly Catholics.

Q. What *universities* can you mention?

A. Those of Leyden, (La'dn) Utrecht, (U'tret) Gron'ingen, and Louvain, (Loo-vane') those of Leyden and Louvain are the most celebrated.

Q. What is the *character* of the Dutch?

A. They are neat, frugal, industrious, cool, and phlegmatic.

Q. What can you say of the *canals?*

A. They are very numerous. The travelling is generally done on them in covered boats, called Treck-Shuits, or *drag-boats*, drawn by horses.

Q. Which are the principal *productions?*

A. Corn, flour, butter, and cheese.

Q. What is the *government?*

A. A limited, hereditary monarchy.

Questions.—Map of Europe.

How is the Netherlands bounded N. E. S. W.? What rivers pass through it? Into what sea do they enter? What is the capital? How situated? How are Brussels and the Hague situated?

Germany.

Peasants. *Bavarians.*

Q. How is Germany *bounded?*

A. N. by the N... S..., D..., and the B... ; E. by P. ., and the A... D... ; S. by the A... D..., and S... ; W. by F... and the N...

1. The *northern part* of Germany is flat; the *southern part* is diversified with plains and ranges of mountains.

2. Much of the *soil* is fertile; yet sandy plains and barren heaths abound on the northwest, and swamps and marshes on the northeast.

3. The *climate* is generally temperate and salubrious. The winters of the northern parts are sometimes severe.

4. *Length*, 650 miles; *breadth*, 600; *square miles*, 225,000. *Pop.* 30,000,000.

Q. Which are the principal *mountains?*

A. The Erzeberg and Carpath'ian.

Q. Which are the principal *rivers?*

A. The Oder, Elbe, Weser, Rhine, Maine, and Danube.

The *Oder rises* in Moravia, and *flowing* northwesterly, *falls* into the Baltic. The *Elbe rises* in Bohemia, and *running* in a northwest direction, *flows* into the North Sea; the *Weser flows* northwest and *falls* south of the Elbe; the *Rhine rises* in Switzerland, and *running* westerly, and again northerly, *divides* Germany from Switzerland and France, whence *passing* through a part of the Prussian dominions, it *enters the* Netherlands, and *flowing* westerly, *discharges* itself into the North Sea. The *Maine* is a branch of the Rhine.

13

It *rises* in Bavaria, and in general has a westerly *direction.* The *Danube rises* in Baden, the southwest corner of Germany, and *flowing* at first northeast, and then southeast, and afterwards in an easterly direction, *passes* through the Austrian dominions and Turkey, and *flows* into the Black Sea.

Q. Which are the principal *religious denominations* ?

A. Protestants and Roman Catholics. The Protestants live in the north, and the Catholics in the south of Germany. They are about equally divided.

Q. What *universities* does Germany contain ?

A. They are about 20 in number, and contain 9000 students; the most celebrated are Got'tingen, Leipsic, (Lipe'sic) Hal'le, and Jan'na.

The *Germans excel* in critical learning, mathematics, philosophy, and mechanics. The *number of books published* in Germany is much greater than in any other country. Authorship is extensively made a business for life.

Q. What is the *character* of the Germans ?

A. They are intelligent, grave, industrious, and persevering.

Q. How is Germany *divided?*

A. Into Austrian dominions in Germany; Prussian dominions in Germany; Holstein and Lauerburg, belonging to Denmark; Luxemburg, belonging to Netherlands; the kingdoms of Bavaria, of Wirtemburg, of Hanover, of Saxony, Grand Duchy of Ba'den, besides 23 smaller states, and four free cities, viz. Lubeck, Frank'fort on the Maine, Bre'men, and Ham'burg.

Q. What is the *government* of these states or kingdoms ?

A. It is generally an absolute, despotic monarchy

The *German states*, although independent, *are united for* the purposes of mutual defence and protection. They have a *Federative Diet*, consisting of 17 plenipotentiaries, and a *General Assembly*, consisting of 69 members.

Q. Which is the *capital* of Germany ?

A. F...t, on the Maine, is the place where the Federative Diet and the General Assembly meet.

1. AUSTRIAN DOMINIONS IN GERMANY.

For an account of these dominions, see Austria.

2. PRUSSIAN DOMINIONS IN GERMANY.

See Prussia.

3. KINGDOM OF BAVARIA.

Q. Where is Bavaria *situated?*

A. In the southeast part of Germany.

Q. Which is the *chief town ?* .

A. Munich, (Mu'nick.)

1. *Munich, 4th class, is on* the Iser, 200 miles west of Vienna; it is one of the most pleasant cities in Europe.

2. The *other principal towns* are Augsburg, *6th class,* northwest of Munich. Ratisbon, *6th class,* on the Danube, northeast from Munich.

4. KINGDOM OF WIRTEMBURG.

Q. How is Wirtemburg *situated ?*

A. In the southwest of Germany, between Bavaria in the east, and Baden on the west.

Q. Which is the *chief town ?*

A. Stuttgard.

Stuttgard, 6th class, is not remarkable either for situation, or appearance. It has a large library, an university, and a magnificent palace.

5. KINGDOM OF HANOVER.

Q. How is Hanover *situated ?*

A. In the north of Germany.

Q. *To whom* does this kingdom *belong ?*

A. To the king of Great Britain, who governs it by a council of regency.

Q. Which is the *chief town ?*

A. Hanover.

1. *Hanover, 5th class,* is on the Soine. Its palace is magnificent.

2. The *other principal town* is Gottingen, *large town,* 60 miles south of Hanover, and *contains* one of the most celebrated universities in the world; it has, besides 66 *professors,* and 1000 *students,* a *library* of 200,000 volumes.

6. KINGDOM OF SAXONY.

Q. How is Saxony *situated ?*

A. In the centre of Germany.

Q. Which is the *chief town ?*

A. Dresden.

1 *Dresden, 4th class,* is on the Elbe, and is noted for its

collections of the fine arts, and for its manufacture of porcelain.

2. *Leipsic*, (Lipesic) 5*th class*, is west of Dresden, and is celebrated *for* its university, and for the fairs, which are held here three times a year, at which great numbers of books are bought and sold.

7. GRAND DUCHY OF BADEN.

Q. How is Baden *situated?*

A. In the southwest corner of Germany.

Q. Which is the *chief town*, and which the *capital?*

A. Manheim is the chief town; Carlsruhe, (Carlsroo'e) the capital.

1. *Manheim*, 6*th class*, is one of the most elegant towns in Europe.

2. *Carlsruhe*, 6*th class*, is a handsome town, built principally of stone. Its streets are regular, and in form the city resembles an open fan.

8. SMALLER STATES OF GERMANY

Q. How are the 23 smaller states *situated?*

A. Principally in the northern and middle parts of Germany.

Q. What *general remark* can you make respecting them?

A. They are small, and unimportant in extent and resources; containing each, from about 5000 to 200,000 inhabitants, and are governed by independent princes.

9. FREE CITIES.

Q. Which are the *free cities* of Germany?

A. Hamburg, Lubeck, Frankfort on the Maine, and Bremen.

1. *Hamburg*, 3*d class*, *lies on* the Elbe, 70 miles from its mouth. It is one of the most *celebrated* commercial cities of Europe.

2. *Lubeck*, 5*th class, is northeast of* Hamburgh, 8 miles from the Baltic. Its trade is checked by its vicinity to Hamburg.

3. *Frankfort*, 4*th class, is on* the Maine, and is the permanent seat of the Diet; it is one of the principal cities of Europe for the extent and amount of its commerce

4. *Bremen*, 5*th class*, is 54 miles southwest of Hamburg; it is generally well built; and has an extensive commerce.

Questions.—Map of Europe.

How is Germany bounded N. E. S. W.? What rivers pass through rt ? What mountains lie in Germany ? What is the capital ? *In what latitude ? Which way from Frankfort is Hanover ? Where is Hamburg ? Bremen ?*

Austrian Empire.

View of Vienna.

Q. How is Austria *bounded?*

A. N. by s..., G..., P..., and R... ; E. by R... and T... ; S. by T...., G... of v..., and I... ; W. by I..., s..., and G...

1. The *German part of Austria*, is generally mountainous. *Hungary, Galicia,* and *Lombardy,* contain extensive plains.

2. The *climate* is mild and healthful, and much of the *soil* is productive.

Q. How is Austria *divided?*

A. Into the German States ; Austrian Poland, or Galicia, Hungarian States, and Austrian Italy.

			Population.
		German States,	9,200,000.
		Austrian Poland,	3,800,000.
Sq. miles of the empire,	270,000.	Hungarian States,	11,000,000.
		Austrian Italy,	4,014,000.

Total 28,014,000.

Q. Which are the principal *mountains?*

A. The Alps and the Carpath'ian mountains.

The *Alps separate* Austrian Italy from the west of the empire ; the *Carpathian separate* Hungary from Galicia.

13*

Q. Which is the principal *river*?

A. The Danube.

1. The *Danube*, *2d class*, *rises* in the southwest corner of Germany, through which *flowing* easterly, it passes into Hungary, where it *turns* south, then southeast, and *passing* into Turkey, it *flows* into the Black Sea.

2. The *Drave* and *Save*, *4th class*, are branches of the Danube.

3. The *Po*, *4th class*, is the great river of Austrian Italy It *falls* into the Gulf of Venice.

Q. Which is the *chief town* and *capital*?

A. V...a.

1. *V...a*, *2d class*, is on the Danube. It *holds* a high *rank among the cities of Europe*, in science, arts, commerce, and refinement.

2. The *principal towns* in several of the divisions of Austria, are Prague, *4th class*, the *capital* of Bohemia; Lemburg, *4th class*, *capital* of Galicia; Gratz, *5th class*, *capital* of Stiria; Buda, *5th class*, *capital* of Hungary; and Hermanstadt, *6th class*, *capital* of Transylvania.

Q. What *universities* does Austria contain?

A. Those of Vienna, Prague, Pest, and Lemberg.

Q. Which is the prevailing *religion*?

A. The Roman Catholic is the established religion; but all others are tolerated.

Q. What can you say of its *agriculture, commerce,* and *manufactures*?

A. Agriculture is the chief business of the people, but is not well conducted; the commerce of Austria is small; manufactures are imperfect, and insufficient for home consumption.

Q What is the *character* of the Austrians?

A. Their character is various. Austria Proper contains a nobility, haughty, ignorant, and vicious, but the people are moral.

Q. Which are the principal *minerals*?

A. Gold, silver, copper, coal, and salt.

Austria is richer in minerals than any other country in Europe. The *principal mines are in* Hungary, and Transylvania, on the borders of the Carpathian mountains. The salt mines of Wielicza, a town of Galicia, 8 miles from Cracow, are *the most celebrated.* The principal mine is 2000 feet *broad,* and 800 *deep.* Many of the persons employed here, were born in the mine, and never go out. It has already been *worked* 600 *years,* and is apparently inexhaustible.

Q. What is the *government?*

A. A hereditary monarchy, limited in Hungary and Transylvania, by constitutional provisions.

Questions.—Map of Europe.

How is Austria bounded N. E. S. W.? What mountains are in Austria? What mountains separate Austria from Italy? What large river passes through Austria, *and in what direction?* What are its branches? What is the capital of Austria? *How situated? In what latitude? Which way from Vienna is Presburg? Buda?*

Prussia.

View of Berlin.

Q. What can you say of the *Prussian Dominions?*

A. They consist of two territories entirely distinct and separate from each other, one lying in the east, and the other in the west, of Germany.

Q. Which division is the *largest,* and how is it *bounded?*

A. The eastern division, consisting of seven provinces, is much the largest, and is bounded N. by the B...; E. by R...; S. by R..., the A... D..., and G...; W. by several small G... s...

Q. How is the western division *situated?*

A. It contains only three provinces, and lies on both sides of the Rhine; it is bounded west by the Netherlands, and on all other sides by Germany.

1. Prussia is generally a *level country*. The *soil varies* from being very barren to that which is fertile.

2. The *climate* of the western parts of Prussia is mild, the eastern parts are cold on account of the winds from Russia.

3. *Length*, 800; *breadth*, from 70 to 300; *square miles*, 108,000. *Pop.* 10,500,000.

Q. Which are the principal *rivers?*

A. The Vistula, Oder, Elbe, and Rhine.

1. The *Vistula rises* in the south of Prussia, and *flowing* generally in a northerly direction, *falls* into the Baltic. The *Oder rises* near the source of the Vistula, and *flowing* in a northwest direction, *enters* the Baltic.

2. The *Elbe rises* in the Austrian dominions, and *flows* by a northwest *course* into the North Sea. The *Rhine passes* through the heart of the great western division of Prussia, from southeast to northwest.

Q. Which is the *chief town* and *capital?*

A. B...n.

1. *B...n, 3d class*, is a beautiful city *on the* Spree, a branch of the Hovel, which is itself a branch of the Elbe; it is *distinguished* for the extent of its manufactures, and the number of its magnificent edifices, and well endowed literary institutions.

2. The *other principal towns* in Eastern Prussia, are Breslaw, *4th class*, in Silesia; Koningsberg, *4th class*, in the eastern part of the kingdom; and Dantzic, *4th class*, on the Vistula, a place of considerable trade.

3. In the *western division*, are Cologne (Co-lo′ne) *5th class*, and Aix la Chapelle (Aix-lah-shap′pel) *5th class*, which is *distinguished* for two celebrated treaties concluded in it, and for its warm baths; also Dusseldorf and Coblentz, *6th class*.

Q. What *universities* does Prussia contain?

A. Those of Halle, Koningsberg, and Berlin.

Q. What is the *character* of the Prussians?

A. The Prussian dominions embrace a variety of character—in general the Prussians resemble the Germans, though they are less intelligent.

Q. What is the prevailing *religion?*

A. The Lutheran is the established religion; but all others are tolerated.

Q. What is the *government?*

A. An absolute monarchy.

Questions.—Map of Europe.

How is Prussia situated ? (See Geog. p. 151.) What three rivers empty into Prussia ? What large river passes through it ? Which is the capital ? *How situated ? Where are Dantzic and Koningsberg ?*

Russian Empire.

Q. Of how many *parts* does the Russian Empire *consist ?*

A. Two, viz. Russia in Europe, and Russia in Asia.

1. The *Russian Empire* is the largest in the world, containing between 6 and 8 millions of *square miles.*

2. The *population* of the Russian empire is estimated at 50 millions.

Q. Which is the prevailing *religion ?*

A. The established religion of Russia is that of the Greek church ; but all others are tolerated.

Q. What is the *character* of the Russians ?

A. The *European* Russians are in general very illiterate, rude, and barbarous, but very brave ; the *Asiatic* Russians consist principally of native, barbarous tribes.

Q. What is the *government* of Russia ?

A. A limited monarchy.

RUSSIA IN EUROPE.

View of Petersburg.

Q. How is Russia in Europe *bounded ?*

A. N. by the F... o... ; E. by R... in A... ; S. by

B... in A..., B. S..., T..., and A...; W. by P..., the B... S..., G... of B..., S..., and N...

1. The *face of the country* is principally level, abounding with vast plains, called *steppes*, forests, and morasses.

2. The *soil* in the *northern provinces* is barren, but the *southern provinces* are as fertile as other countries in the same latitude.

3. The *climate* in the *south* is warm, but the *northern parts* are extremely cold. The *method of warming houses in Russia* is by an oven, constructed with several flues, and they regulate the warmth in the apartments by a thermometer, with great exactness, opening and shutting the flues to increase or diminish the heat.

4. *Square miles*, 1,500,000. *Pop.* 40,000,000.

Q. Which are the principal *mountains?*

A. The Ural mountains.

This chain is 1500 miles *long*, and *separates* Europe from Asia. The *highest peaks* are of the *5th class.*

Q. What *seas* border upon Russia?

A. The White Sea, in the north; the Baltic, in the west; the Black Sea, and Sea of Azof in the south.

Q. Which are the principal *bays* or *gulfs?*

A. The Gulfs of Both'nia, Fin'land, and Ri'ga.

Q. Which are the principal *lakes?*

A. Lake Lado'ga and One'ga.

1. *Lake Ladoga lies* east of the Gulf of Finland, and is one of the largest lakes in Europe, being 140 miles *long*, and 75 *broad*.

2. *Onega lies* east of Ladoga, and is 130 miles *long*, and 70 *broad*.

Q. Which are the principal *rivers?*

A. The Vol'ga, Don, Dnieper, Dniester, and Dwi'na.

1. The *Volga*, 3d *class*, *rises* between Petersburg and Moscow, *runs* east and south, and *falls* into the Caspian Sea, by more than 70 mouths.

2. The *Don*, 3d *class*, *rises* south of Moscow, *runs* south, and *flows* into the sea of Azof.

3. The *Dnieper*, 2d *class*, *rises* west of Moscow, and *running* south, *falls* into the Black Sea.

4. The *Dniester*, 3d *class*, *rises* in the Carpathian mountains, *runs* southeast, and *flows* into the Black Sea.

5. The *Dwina*, 3d *class*, *rises* near the source of the Volga and *flows* west into the Gulf of Riga.

Q. Which is the *chief town* and *capital?*

A. St. P...g.

1. *St. P...g, 2d class,* was *founded* by Peter the Great, in 1703, and is now one of the most splendid cities in the world. It *stands* on the Neva, near its entrance into the Gulf of Finland.

2. *Moscow, 2d class,* was the ancient capital; it is *situated* near the centre of Russia in Europe; and previous to its being burned by order of the Russian governor, in 1812, to prevent the French from deriving any advantage from possessing it, contained 300,000 inhabitants. It has been in a great measure re-built.

3. The *other principal towns* are Cronstadt and Riga, on the Baltic; Odes'sa on the Black Sea, *5th class,* and Archangel, (Ark'angel) *large town,* on the White Sea.

Q. What *universities* does Russia contain?

A. Universities at St. Petersburg, Warsaw, Abo, and at other places.

But little *attention has been paid to education,* in Russia, until lately. Much encouragement, however, *has recently been given to learning* by the late Emperor Alexander, and many of the Russian youth, in the higher classes, are now well educated.

Q. What is the *character* of the Russians in Europe?

A. They are illiterate, rude, and barbarous; but very brave.

Q. What can you say of the *manufactures* of Russia?

A. They manufacture linen, silk, paper, carpets, hats, and the leather which takes its name from the country.

Q. Which are the principal *productions* and *exports?*

A. Furs, peltry, and hemp.

Q. Which is the principal *island* belonging to Russia?

A. Nova Zembla.

1. *This island lies* N. E. of Archangel; it is 300 miles *long,* and from 100 to 200 *broad.* It is *inhabited* only by white bears, white foxes, rein deer, and rabbits; on the *coast are found* whales, sea dogs, sea cows, and sea lions.

2. The *other islands* are Oesel, Dago, and Aland, in the Baltic; and Spitzber'gen, in the Northern Ocean.

Questions.—Map of Europe.

How is Russia in Europe bounded N. E. S. W.? What rivers empty into the Baltic sea? What into the Black sea? What river empties into the Sea of Azof? What river rises in R. in Europe and flows into the Caspian Sea? *Where do these rivers rise, and what is their course?* What sea is in the north of R..? What mountains separate R. in Europe from R. in Asia? What lakes does R. contain? With what gulf are they connected? What gulf lies south of the Gulf of Finland? What islands between these gulfs? What sea lies north of the Black Sea? What is the peninsula called which lies north of the Black Sea? Which is the chief town and capital? *On what gulf is it situated? In what latitude? Where is Archangel? Moscow? Riga? Smolensk? Cherson? and Odessa?*

Poland.

Q. How is Poland *bounded?*

A. N. and E. by R...; S. by R... and A...; W. by A..., P... and the B...

1. *Poland* was *formerly* an independent, and powerful kingdom, *comprehending* a much larger extent of territory than it now does.

2. *In* 1772, the country being distracted by internal dissensions, *Russia, Prussia, and Austria*, took possession of a large part of it; and divided it among themselves. *In* 1793 they made a further division, and in 1795, they *annihilated* the kingdom.

3. The *portion* seized by Russia was *erected into a kingdom* in 1815, of which the Emperor of Russia *was made king*.

4. The *face of the country* is in general level.

5. The *climate* is cold, but the *soil* productive.

6. *Square miles*, 47,000; *Pop.* 2,800,000.

Q. Which is the principal *river?*

A. The Vistula.

The Vistula is the *great river of Poland;* it *rises* at the foot of the Carpathian mountains, *passes* by Cracow, Warsaw, and Thorn, and generally *flowing* northwesterly enters the Baltic, at Dantzic.

Q. Which is the *chief town* and *capital?*

A. W...w.

1. *W...w, 4th class*, is *situated* on the Vistula; it is surrounded by a moat, and double wall. It exhibits a great contrast of wealth and poverty, luxury and distress.

2. *Cracow, 5th class*, on the Vistula, is in the southwest part of Poland. It is a free city.

3. *Four of the other Polish towns*, are Dantzic, *4th class*, and Posen, *6th class*, belonging to Prussia; Lemberg, *5th class*, to Austria, and Wilna, *6th class*, to Russia.

Q What is the *character* of the Poles?

A. They are one of the most rude and illiterate nations of Europe.

Q. What is the *religion* of the Poles?

A. Roman Catholic.

Poland is said to contain more *Jews* than all the rest of Europe. The trade is chiefly in their hands.

Q. What celebrated *mines* does Poland contain?

A. Salt mines. *(See Austria.)*

Questions.—Map of Europe.

How is Poland bounded N. E. S. W.? *Where is Warsaw, the capital?* What is the great river of Poland? *Where does it rise? What is its course? Where does it empty? Where is Cracow?*

Denmark.

View of Copenhagen.

Q. How is Denmark *bounded?*

A. N. by the ST... of s... R..., and c..., which separate it from N..., and s...; E. by the c..., and the B...; S. by G...; W. by the N... s...

1. The *kingdom of Denmark consists of* the peninsula of Jutland, the islands of Zea'land and Fu'nen, and others at the entrance of the Baltic, together with Iceland in the Atlantic, and the Faroe islands, between Iceland and Norway.

2. The *climate* of the island of Zealand, and the south of Jutland, is rendered temperate by vapours from the sea, but the north of Denmark is a cold country.

3. The *soil* of the southern part is fertile, but less fertile in the north.

4. *Square miles*, 21,615; *Pop.* 1,500,000.

14

Q. Which is the *chief town* and *capital?*

A. C...n.

1. *C...n, 3d class, stands* on the island of Zea'land, **and is** one of the best built cities in Europe. It has a fine harbour and extensive commerce.

2. The *other principal towns* are Altona, *6th class*, on the Elbe, and Elsinore, (El-se-nore') *large town,* 20 miles north of Copenhagen, a well known seaport, at which all the foreign ships which trade to the Baltic, pay toll.

Q. What *universities* does Denmark contain?

A. One at Copenhagen, and another at Kiel, (Keel.)

The Danes are very attentive to the *education of their children,* and the common schools are very respectable.

Q. Which are the principal *productions* and *exports?*

A. The *productions* are grain, beans, and potatoes; the *exports,* grain, cattle, horses, beef, and pork.

Q. What is the *character* of the Danes?

A. They are honest and industrious, but timid, and much addicted to intemperance.

Q. What is the *government?*

A. An absolute hereditary monarchy.

Q. What large *island* in the Northern Ocean belongs to Denmark?

A. Iceland, famous for its volcano, Mount Hecla; and for its hot springs, called Geysers.

1. *Iceland* is about 280 miles *long,* and 210 *broad,* containing 40,000 *square miles.* It is a cold, barren, and thinly inhabited region, *containing* between 50 and 60,000 *inhabitants,* who enjoy few of the comforts of life, and *subsist almost wholly on* animal food; chiefly fish.

2. *Mount Hecla, 6th class,* is one of the most celebrated volcanoes in the world; it has numerous and dreadful eruptions. Matter is said to have been thrown from it 150 miles.

3. The *Geysers,* or Hot Springs, are great curiosities.—They throw into the air jets of boiling water, sometimes to the height of 200 feet, with a noise like that of cannon.

Q. What is the *character* of the Icelanders?

A. They are simple, intelligent, moral, and religious.

The *religion of the Icelanders* is Lutheran. The *number of churches* exceeds 300.

Questions.—Map of Europe.
How is Denmark bounded N. E. S. W. ? What is the capital ?—
How situated ? Where is Elsinore ? Altona ? Where are the
Faroe islands ? Where is Iceland, which belongs to Denmark ?
What volcano is on this island ?

𝕾𝔴𝔢𝔡𝔢𝔫.

View of Stockholm.

Q. How is Sweden *bounded ?*

A. N. by N... ; E. by R..., G. of B..., and the B...;
S. by the B... ; W. by the C..., and N...

1. The *face of the country* is in general mountainous, interspersed, however, with level tracts, and surrounded by
ranges of mountains, of which the Dofrafield range between
Sweden and Norway, is the principal.

2. The *climate* in winter is very cold ; the *summers* short
and hot.   The *transition from the one to the other* is so
sudden, that spring and autumn are scarcely known.

3. The *soil* of Sweden is poor ; the southern parts are the
best and the most populous.

Length, 1000 miles ; *breadth,* from 200 to 300 ; *Pop.*
2,500,000.

Q. Which are the principal *lakes ?*

A. Wenner, Wetter, and Malar.

The lakes of Sweden are numerous.   Those mentioned are
the largest, and of them Wenner is the principal.

Q. Which are the principal *rivers ?*

A. Gotha, Dal, and Motala.

1. The *Gotha*, *5th class*, whose *course* is south of west, is the *outlet* of lake Wenner, connecting it with the Cat'tegat.

2. The *Dal*, *5th class*, *rises* in the mountains on the borders of Norway, and after a very circuitous *course*, *empties itself* 60 miles north of Upsal.

3. The *Motala*, *6th class*, is the *outlet* of lake Wetter ; it *flows* in an easterly direction, and *falls* into the Baltic.

Q. Which is the *chief town* and *capital?*

A. S...m.

1. *S...m*, *4th class*, is *built on* several rocky islands, which are united by wooden bridges It *stands* at the junction of the lake Malar with the Baltic. The royal palace is magnifi. cent.

2. *Got'tenburg*, *6th class*, on *the* Cattegat, is in the south-west part of Sweden, and has an extensive trade in iron.

3. The *other principal towns* are Carlscro'na; *6th class*, *noted* for being the chief station of the Swedish navy, and Up'sal, *large town*, for its university.

Q. What *universities* does Sweden contain?

A. Those of Upsal and Lund, the former of which is one of the most distinguished in Europe.

The Swedes pay much *attention to education*, and have cultivated the sciences with much success.

Q. Which is the prevailing *religious denomination?*

A. The Lutheran.

Q. What is the *character* of the Swedes

A. They are intelligent, lively, honest, hospitable, and temperate.

Q. Which are the principal *productions* and *exports?*

A. Iron, copper, alum, timber, and tar.

The Swedish iron is esteemed the best in Europe. *The iron mine most celebrated is* that of Dennemora, 60 miles N. W. of Stockholm.

Q. What is the *government?*

A. A limited, hereditary monarchy.

Questions.—Map of Europe.

How is Sweden bounded N. E. S. W.? What mountains separate it from Norway? What river from Russia? What 3 lakes do you notice in Sweden? What is the chief river? What is the capital? *How situated? Where is Upsal? Carlscrona? Gottenburg?*

Norway.

Maelstroom.

Q. How is Norway *bounded?*

A. N. by the F... O...; E. by R..., and S...; S by S..., the S..., and N... S...; W. by the A... O...

1. *This country, next to Switzerland,* is the *most mountainous* of any country in Europe, and abounds with bold and picturesque scenery,

2. The *climate* is extremely cold in winter; the *summers* are short and hot.

3. The *soil* is generally poor, but better adapted for pasturage, than for grain.

4. *Length,* 350 miles *; breadth,* 250; *square miles,* 161,000. *Pop.* 1,000,000.

Q. What can you say of the *rivers?*

A. They are numerous, but inconsiderable in length.

Q. What *capes* can you mention?

A. North Cape, and Naze.

Q. Which is the *chief town,* and which the *capital?*

A. B...n is the chief town; C...a, the seat of government.

1. *B...n, 6th class,* has an excellent harbour, and considerable trade.

2. *C...a, 6th class,* in the south, contains a university, and has a flourishing trade.

Q. Which is the prevailing *religion?*

A. Protestantism.

14*

Q. What celebrated *whirlpool* lies near the coast ?

A. The Maelstroom (Mawl'stroom.)

This whirlpool is *supposed to be occasioned by* the rapid ebb and flood of the sea, between the island Moskoe, and a neighbouring island. *Its influence is often felt at the distance of* nine miles, and sometimes draws in, and overwhelms ships. and whales, when at three miles distance.

Q. What is the *character* of the Norwegians ?

A. They are temperate, rude, and illiterate.

Q. Which are the principal *exports* ?

A. Timber, fish, copper, and silver.

Q. What is the *government* ?

A. Monarchical.

Since 1814, *Norway* has been *subject to* Sweden, but it has enjoyed its own legislature, and a separate administration.

Questions.—Map of Europe.

How is Norway bounded N. E. S. W. ? What are the northern and southern capes ? What straits separate it from Sweden ? Where is the Maelstroom ? What islands on the N. W. coast ? What is the capital ? *How situated ? Where is Drontheim ? Christiana ?*

Lapland.

Reindeer.

Q. How is Lapland *situated* ?

A. In the northwest of Europe, having the Frozen ocean north ; the White Sea east ; the Gulf of Bothnia south ; and the Atlantic west.

1. The *eastern part* is called Russian Lapland ; the *middle,* Swedish Lapland ; and the *western,* Norwegian Lapland. The *country belongs* to Russia, Sweden, and Norway.

2. The *face of the country* is mountainous, interspersed with numerous lakes and ponds.

3. The *winters* are intensely cold. In some parts of Lapland the *sun in winter is absent* several weeks.—The *nights,* however, are *rendered less dreary* than might be expected, *by* the moon, stars, and northern lights, which are almost constantly visible. In *summer, the sun does not set for* the same length of time, in consequence of which the heat becomes excessive.

4. The *population* is estimated at 60,000.

Q. What is the *character* of the Laplanders?

A. They are dwarfish, ignorant, superstitious, and barbarous.

The Laplanders are about four feet *high.* They *live in* small huts, in the centre of which a fire is built, around which they sit upon their heels. In this posture, they eat their food, which generally *consists of* reindeer. *In winter, they perform journeys on* sledges, which the reindeer will draw 200 miles in a day.

Questions.—Map of Europe.

How is Lapland bounded N. E. S. W.?

ASIA.

Turks.   *Persians.*

Chinese.   *Hindoos.*

Q. How is Asia *bounded?*

A. N. by the r... o...; E. by the p...; S. by the

I... o...; W. by A..., from which it is separated by
the R... s..., the M... s..., and E...

1. *Asia* is the largest and most populous division of the
world, and *has been the scene of* the most important events,
which history records. *There*, the first man was created;
the patriarchs lived; the law was given to Moses; the first,
and greatest, and most celebrated monarchies were formed; and thence the founders of the first cities and kingdoms
in other parts of the world, led forth their colonies. In Asia
too, the Son of God appeared, suffered, and died; and thence
began the light of that gospel, which is ultimately to bless all
nations.

2. Asia has every variety of *climate*, from the intense heat
of the torrid zone, to the extreme cold of the arctic, or frozen regions.

3. The *greatest length* of Asia, taken obliquely from the
Isthmus of Su'ez to Behring's strait, is about 7370 miles;
the *breadth*, from cape Taymour, in Siberia, to cape Comorin,
in India, is about 4230. Asia is estimated to contain nearly
16,000,000 *square miles*.

Q. Which are the principal *mountains?*

A. The ranges of Al'tay, Him'maleh, Be'lur Tag,
and China.

1. The *Altaian* range, highest peaks, *4th class*, is the most
noted. It *commences* near the Sea of Aral, not far from the
Ural range, (*see mountains of Europe,*) and *runs* northeast
to the Pacific ocean, between Russia in Asia and the Chinese
empire, taking different *names in its progress*. *This chain* is
supposed to be 5000 miles *long*.

2. The *mountains of Himmaleh*, Da-wa-la-ge'ri, the highest
peak, *1st class*, *separate* Hindoostan from Thibet and Tartary.
They *commence* in north lat. about 35°, and *run* in a southerly
direction from Persia, towards the southern borders of China,
to north lat. 25°.

3. The *Be'lur Tag*, or Cloudy Mountains, *run* from north
to south, *connecting* the western extremities of the Altaian
and Himmaleh ranges.

4. The *mountains of China connect* the eastern extremities.

5. Hence it appears that *Thi'bet and Tar'tary are encircled*
by a chain, or belt of mountains—the Altaian range being
on the *north;* the mountains of Him'maleh on the *south;*
the Belur Tag on the *west;* and the mountains of China on
the *east*. These countries are supposed to be in general about
9000 feet above the level of the sea, and are called the table
land of Asia. (*See Tartary.*)

6. Besides the above *mountains may be mentioned* the
mountains of Cau'casus, a range 400 miles in *extent*, highest

peaks, *5th class*, between the Black Sea and the Caspian ; the mountains of Tau'rus near the shore of Anto'lia (in Turkey) east of Rhodes ; Mount Leb'anon, *5th class*, in Syria, celebrated in ancient times for its excellent cedars ; Mount Ar'arat, *5th class*, in Arme'nia, near Cau'casus ; the Gauts, *6th class*, a chain of mountains parallel with the Malebar coast of Hindoostan.

Q. Which are the principal *seas ?*

A. The sea of Kamtschat'ka, of Ochotsk', of Japan, the Yellow and China seas on the *east.* The Arabian sea on the *south;* the Caspian Sea, and Sea of Aral in the *west.*

The *Caspian Sea* and *Sea of A'ral* are more properly lakes. The former *lies* east of the Black Sea. Its water is salt, but it has no *visible* communication with the ocean, and no outlet. It is 2400 miles in *circumference.* It is supposed by some that it communicates with the Indian ocean by some subterraneous passage. The *Sea of Aral lies* east of the Caspian, and is 200 miles *long.*

Q. Which is the principal *lake ?*

A. Baikal', in Siberia, north of the Altaian mountains.

Baikal is the largest fresh water lake in Asia. Its *length* is 360 miles ; *breadth* from 20 to 50 miles. It *contains* about 2000 *square miles more than lake Erie.*

Q. Which are the principal *straits ?*

A. Babelman'del, Or'mus, Malac'ca, Sun'da, and Behring's strait.

1. The strait of *Babelmandel* is between Asia and Africa.

2. The strait of *Ormus* is at the entrance of the Persian gulf.

3. The strait of *Malacca* is between the Malay coast and the island of Suma'tra.

4. The strait of *Sunda* is between Suma'tra and Java.

5. *Behring's* strait is between Asia and America.

Q. Which are the principal *bays* or *gulfs ?*

A. The Persian gulf, Bay of Bengal, gulfs of Siam, Tonquin, Corea, and Obi.

1. The *Persian gulf* is between Persia and Arabia.

2. The *Bay of Bengal* is between Hindoostan and the Birman empire.

3. The *Gulf of Siam* and *Tonquin* are arms of the Chinese Sea.

4. The *gulf of Corea* is between Corea and Niphon, one of the Japan Isles.

5. The *gulf of Obi* is in the northwest, and communicates with the Frozen ocean.

Q. Which are the principal *rivers*?

A. The Obi, Yenisei, (or Jenesea) and Lena, on the *north ;* the Amour, Hoang-Ho, Kiang-Ku, in the *east ;* and the Ganges, Indus, and Euphrates, in the *south* and *west*.

1. The *Obi, Yenisei,* and *Lena, 1st class, all rise* in the mountains of Altay, and *flowing* north, *fall* into the Frozen Ocean.

2. The *Amour, Hoang-Ho,* and *Kiang-Ku, 1st class, all rise* south of the Altay mountains ; the *Amour flows* east and *falls* into the Sea of Ochotsk ; the *others empty* themselves into the sea east of China ; the *former winding* to the north, and the *latter* to the south.

3. The *Ganges, 2d class, rises* in the Himmaleh mountains, and *flows* southeast into the bay of Bengal. The *Indus, 2d class,* also *rises* in the Himmaleh mountains, and *flows* southwest into the Indian ocean. The *Euphrates, 2d class, rises* near the Black Sea, and *flowing* southeast, *falls* into the Persian gulf.

Q. Where is the isthmus of *Su'ez* ?

A. Between the Mediterranean and Red Sea. It connects Asia with Africa.

Q. Which are the principal *capes*?

A. Cape Tay'mour, in the Northern Ocean; Cape Com'orin, in the Indian Ocean ; and Cape Lopat'ka, in the Pacific.

Q. What is the prevailing *religion* of Asia ?

A. Paganism and Mahometanism.

Q. What *countries* does Asia contain ? .

A. Turkey in Asia, Russia in Asia, Arabia, Persia, Independent Tartary, Hindoostan, India beyond the Ganges, the Chinese Empire, and Japan.

Questions—Map of Asia.

How is Asia separated from Europe ? From Africa ? *How far does it extend north?* What strait separates it from America ? What Ocean lies on the east ? How is it bounded on the south ? Which are the principal countries in the northeast ? South ? West ? What chain of mountains lie in the middle of Asia ? What is their direction ? What names does the range take in its course ? What chain south of the Altaian running in the same direction? What chain unites these ranges on the west? What chain in the south of Hindoostan? Where are Mt. Horeb? Mt. Sinai? Mt. Taurus? Mt. Ararat? Mt. Cau'casus? What seas are there on the west of Asia? What isthmus between the Red sea and Mediterranean? What two seas are there E. of the Black sea? What gulfs and bays on the south? What seas on the east? What gulf east of Tartary? What east of Tonquin? What cape on the south of Hindoostan? What south of Kamtschatka? What in Siberia? What strait leads into the Red sea? What into the Persian gulf? What between the Malay coast and Sumatra? What between Sumatra and Java? What are the three great rivers in the north of Asia, beginning at the west? *Mention their source, direction, and place of discharge.* What two branches has the Obi? What are the branches of the Yenisei? Through what lake does one of these branches pass? *Describe the three great rivers which are in the east of Asia.* What river empties into the China sea? What into the gulf of Siam? What three large rivers empty into the Bay of Bengal? What river empties into the Arabian sea? Where do the Euphrates and Tigris empty? What river empties into the Caspian sea? What into the sea of Aral? What great inland seas are in the west of Asia not connected with the ocean? Where is lake Baikal? What island lies north of Asia, and in what ocean? What islands in the sea of Kamtschatka? What three large islands lie east of Tartary? What cluster of small islands east of China? What island east of Canton? What cluster south of Formosa, and east of Tonquin? Which way from the Philippines are Borneo and Celebes? What islands lie south and east of Borneo and Celebes? Where is Sumatra? What small islands lie in the bay of Bengal? What large islands south of Hindoostan? What small group west of this?

𝕮𝖚𝖗𝖐𝖊𝖞 𝖎𝖓 𝕬𝖘𝖎𝖆.

View of Jerusalem.

Q. How is Turkey in Asia *bounded?*

A. N. by the S. of M..., the B... S.. and R. m

A...;   E. by p...;   S. by a...;   W. by the m... s...., and the a...

1. *Turkey in Asia comprehends* the countries of Asia Minor, now called 'Nato'lia, Syr'ia, Arme'nia, and the ancient Mesopota'mia.

2. *Asia Minor* is in the northwest; *Syria* in the south west; *Armenia* in the northeast; *Mesopotamia* in the southeast.

3. The *surface* of the country is beautifully diversified by mountains and valleys.

4. The *soil* is generally good, and in some portions of the country is exceedingly rich, but little attention is paid to agriculture.

5. The *climate* is delightful, and naturally salubrious; but this country is often visited by the plague, which sometimes carries desolation in its progress.

6. *Square miles,* 400,000.   *Pop.* 10,000,000.

Q. Which are the principal *mountains?*

A. The mount Tau'rus range, and mount Leb'a-non.

1. The *Mount Taurus range runs* along the shore of the Archipelago, in Asia Minor, east of Rhodes, and stretches east towards the Caspian Sea.

2. *Mount Leb'anon, 5th class, stretches* along the Mediterranean in Syria, generally from 30 to 40 miles from the shore. *This range was celebrated* in ancient times *for* its excellent cedars, and still bears on its summit some of the descendants of those noble trees.

3. To these may be *added,* Mount Ararat, *5th class,* in Armenia; Hermon, *5th class,* Carmel and Tabor, *6th class,* in Syria.

Q. Which are the principal *rivers?*

A. The Euphra'tes, Ti'gris, and Jor'dan.

1. The *Euphrates, 2d class, rises* in Armenia, near the Black Sea, and *flows* southeast into the Persian Gulf. It is *navigable* for ships of 500 tons to Bassora, 70 miles.

2. The *Tigris, 3d class, rises* in Armenia, and *flowing* southeast, *joins* the Euphrates at Corna, about 60 miles above Bassora.

3 The *Jordan, 6th class,* is a river much spoken of in the Scriptures. It *lies* in Palestine, and *flows* south into the lake Asphaltites, or Dead Sea.

Q. What *lake* or *sea* deserves notice?

A. The Asphaltites, or Dead Sea, which occupies the place where Sodom and Gomorrah, and the other cities of the plain, once stood.

It is about 67 miles from *north to south*, and 27 in its *greatest breadth*. It is *called the Dead Sea*, because it preserves nothing alive, and *Asphaltites*, from Asphaltus, a pitchy substance, which rises on its surface, and occasionally explodes.

Q. Which is the *chief town?*

A. A...o.

1. *A...o*, 2d class, *the capital* of Syria, and the largest city in Asiatic Turkey, is 70 miles from the sea. It was nearly destroyed in 1822, by an earthquake. It was *famous* for its manufactures of silk and cotton.

2. *Damascus*, 2d class, *lies* south of Aleppo; it was once *famous* for manufacturing sword blades of a superior quality, called Damascus steel. The silk cloth called *Damask*, derives its name from this city, as also the plum, called the *damson*.

3. *Smyrna*, 3d class, *is on* the Archipelago, and is the chief port for foreign commerce. It is very *subject to* the plague.

4. *Erzerum*, 3d class, *the capital of* Armenia, is near the Euphrates; it has an extensive trade with Persia and India.

5. *Diarbekir*, 3d class, on the Tigris, is *distinguished* for its manufactures.

6. *Jerusalem*, 5th class, is southwest of Damascus; it has a small portion only of its former magnificence.

Q. What is the prevailing *religion?*

A. Mahometanism is the established religion; but one third of the inhabitants are supposed to be Greek Christians.

The high priest of the Turks is *called* Mufti; he is the oracle who is consulted, and who solves all difficult questions in religion; his *decisions are called* fetfas. Even the Sultan *consults him in all difficult cases*, and promulgates no law, makes no declaration of war, establishes no imposts, without obtaining a fetfas.

Q. Which are the principal *manufactures?*

A. Carpets and leather!

Q. Which are the principal *productions?*

A. Silk, corn, cotton, wine, coffee, myrrh, oranges, and lemons.

Q. What is the *government?*

A. An absolute despotism.

15

Questions.—Map of Asia.

How is Turkey in Asia bounded N. E. S. W. ? Where is **Mt. Tan-rus** ? Mt. Ararat ? What rivers flow into the Persian gulf ? **Which is** the chief town and capital ? *Which is the chief town and capital of Armenia? How are these towns situated? Where is Damascus? Jerusalem? Smyrna?* What island lies near the coast of **Syria** ?

RUSSIA IN ASIA.

Travelling in Siberia.

Q. How is Russia in Asia *bounded ?*

A. N. by the F... O... ; E. by B... strait, and the P... ; S. by the C... E..., I... T..., the C... S..., and T... ; W. by R... in E..., from which it is separated by the river Volga and Ural mountains.

1. This *country is generally known by the name of* Siberia, especially that part of it which borders on Russia in Europe. It is *divided* into two governments, Tobolsk', in the west, and Irkutsk', in the east, each of which, in *extent,* is nearly as large as Europe.

2. The *face of the country* is in general even, being an inclined plane, descending from the Altay mountains on the south, towards the Frozen Ocean.

3. The *climate* is intensely cold during the long winters which prevail; the *summers* are very short and hot. The *north* regions admit of no cultivation ; the *middle,* of very little ; but some districts in the *south,* are capable of producing grain.

4. *Square miles,* 6.000,000. *Pop.* 10,000,000.

Q. Which are the principal *mountains ?*

A. The Altaian, the Ural, and the Caucasian.

1. The *Altaian* mountains, highest peaks, *4th class, commence* near the sea of Aral, not far from the Ural range, and *run* northeast to the Pacific, between the Russian and Chinese empires. They have various names in their courses.

2. The *Ural* mountains, 5*th class*, *divide* Európe from Asiatic Russia.

3. The *Caucasian* mountains, highest peaks, 5*th class*, are between the Black sea and the Caspian.

Q. Which are the principal *rivers?*

A. The Obi, Yenisei, and Lena.

1. All these rivers are of the 1*st class*, and have their *source* in the Altay mountains, and *flow* north into the Frozen Ocean.

2. Besides these may be *mentioned* the Irtish, 1*st class*, which *rises* in the mountains of Tartary, and *flowing* north-east, falls into the Obi; and the river Ural which *rises* in the Ural mountains, and *flows* southerly into the Caspian sea.

Q. Which is the *chief town*, and which the *capital of Siberia?*

A. A...n' is the chief town ; T...k' is the capital of Siberia.

1. *A...n*, 4*th class*, *is situated* at the mouth of the Volga. *It carries on a great trade with* Petersburg and Persia ; great quantities of silk are manufactured in its neighbourhood, and are exported.

2. *T...k*, 6*th class*, on the Irtish, is the *chief town of* Western Siberia, and the centre of the Russian fur trade.

3. *Irkutsk*, 6*th class*, is the *chief town of* Eastern Siberia. It is a splendid and prosperous city, and is the principal seat of the commerce between Russia and China.

Q. What is the prevailing *religion?*

A. The established religion is that of the Greek church, but Mahometanism and Paganism extensively prevail throughout the country.

Q. What is the *character* of the inhabitants?

A. They are chiefly Tartars, who are ignorant, filthy, and roving, and subsist principally upon their horses, oxen, sheep, and goats.

1. The *Cossacks*, who so much distinguished themselves in the late wars of Europe, *were originally* Poles, who settled in Russia, and now have an independent military government. They *reside* on the plains near the Don and Volga. The *Calmucks live* between the Volga and Caspian , they are considerably advanced in civilization.

2. The *Circassians* and *Georgians*, particularly the females, are *celebrated for* their beauty, and are often purchased to adorn the eastern seraglios.

Q. What *minerals* are found here ?

A. Gold, silver, and precious stones are obtained

in great quantities, from the mines in the **Altaian** mountains, and elsewhere.

Q. Which are the principal *productions?*

A. The southwestern provinces produce **grain,** figs, and almonds, but the rigour of the climate **does** not admit of cultivating the soil in Siberia.

Q. What *animals* does Siberia contain?

A. The reindeer, beaver, wolves, foxes, **bears;** besides many others which are hunted for their **furs.**

Q. What is the *government?*

A. A limited monarchy, the country being under the dominion of Russia, but the actual power of the Emperor, over the distant provinces, is small.

Questions.—Map of Asia.

How is Russia in A. bounded N. E. S. W.? What is this country generally called? What cape is on the N.? What islands? What chain of mountains in, and around it? What rivers in the northern part? *Where is their source? What their course and place of discharge?* What lake? What gulf in the north? What peninsula in the east? Where is Behring's strait? Sea of Kamtschatka? Of Ochotsk? Which is the chief town? Which is the capital? *How situated?*

Arabia.

View of Mecca.

Q. How is Arabia *bounded?*

A. N. by T..., and P... G..., which separates it from Persia; E. by P..., and the A... S...; S. by the A... S..., ST. of B..., and R... S...; W. by the R... S..., and the I... of S...

1. The *interior part* of Arabia is an immense waste of sand, scarcely watered by a single stream, and has but here and there a verdant spot. The *edges of the country, on the sea coast*, are better watered and more productive.

2. The *climate* is intensely hot and dry.

3. *Square miles*, 1,000,000 ; *Pop.* 10,000,000.

Q. Which are the principal *mountains?*

A. Mount Sinai and Mount Horeb.

Mount Sinai, 6*th class*, is *near* the north end of the Red sea, and is *famous* as the place, where God gave to Moses the moral law, or ten commandments. *Mount Horeb, 6th class, is* not far distant.

Q. Which is the *chief town* and *capital?*

A. M...a.

1. *M...a*, 5*th class*, the birth place of Ma'homet, is a well built city. It occupies a narrow vale, in the midst of a rocky and barren country, about a day's journey *from the Black Sea*. Vast numbers of pilgrims *resort to it* from every quarter of the Mahomedan world.

2. *Medina, large town*, is 180 *miles north of Mecca*, and *contains* the tomb of Mahomet ; the Mosque, which contains the tomb, is supported by 400 columns of black marble. The tomb is surrounded by 300 silver lamps, which are kept continually burning.

3. *Jidda, small town*, is the seaport of Mecca. *Mocha, small town*, a seaport near the strait of Babelmandel, is famous for its coffee.

Q. What is the prevailing *religion?*

A. Mahometanism.

Q. What is the state of *learning* in Arabia ?

A. During the middle ages, the Arabians excelled in literature and science ; but at present, education is so generally neglected, that few are able either to read or write.

Q. What is the *character* of the inhabitants ?

A. On the east the inhabitants are considerably civilized. The interior is inhabited by the *Bedouins*, or wandering Arabs, who are grave, hospitable, and generous to strangers, who visit them, but ignorant, superstitious, and cruel to travellers.

15*

Q. How do they *travel* in Arabia?

A. Principally on camels.

The *Arabian horses* are much *celebrated for* their swiftness, docility, and hardiness; but the *camel* only can well endure the toil of passing over the great deserts; the camel *carries* between 7 and 800 pounds upon his back, and can travel a week without water.

Q. What do you understand by a *caravan*?

A. A number of merchants, travellers, or pilgrims, who go in company across the country.

These caravans frequently *consist of* thousands. They *travel* in this manner *through* Arabia, Persia, Turkey, Tartary, and Africa, and go armed to defend themselves against the Bedouins, or wandering Arabs. Sometimes, however, the Arabs attack and plunder them, and instances have occurred of whole caravans being buried and destroyed by the clouds of sand, which are swept by the wind across the deserts. At other times, a pestiferous wind, called *simoon*, surprises these caravans, and occasions their instant death.

Q. Which are some of the principal *productions*?

A. Coffee, gum Arabic, myrrh, and frankincense.

Q. What is the form of *government*?

A. It is generally monarchical; the country being in possession of chiefs, who are called sheiks, caliphs, imans, and emirs. The government of some of these chiefs is patriarchal; i. e. like that of a father over his family.

Questions.—*Map of Asia.*

How is Arabia bounded N. E. S. W.? What mountains do you notice? In what part of A. are they? What gulf is on the E.?— What sea on the W.? Which is the chief town and capital?— *What is the seaport of Mecca? Where is Mocha? Which way from Mecca is Medina?* What Arabian island near cape Guardafui?

Persia.

View of Ispahan.

Q. How is Persia *bounded* ?

A. N. by the c... s..., i... т..., and c... e...; E. by c... e..., and h...; S. by the a... s... and a..., from which it is separated by the Persian Gulf; W. by a... and т...

1. Persia *abounds* in mountains, and sandy deserts. It has few rains, and *suffers much for want of* water. The *interior consists chiefly of* an immense dry, salt plain, which is said to be 700 miles *long*.

2. The *climate* of the northern part is cool; that of the southern, hot.

.3. *Square miles*, 480,000 ; *Pop.* 18,000,000.

. Q. How is Persia *divided?*

A. Into west and east Persia, or Persia Proper, and the kingdoms of Cabul, and Beloochistan.

WEST PERSIA, OR PERSIA PROPER.

Q. Which is the *chief town,* and which the *capital?*

A. I...n is the chief town ; T...n the capital.

1. *I...n, 2d class,* was the ancient capital ; it was formerly of immense size, and very splendid in its buildings. It is *situated* about half way between the Caspian Sea and Persian Gulf.

2. *T...n, 4th class,* is a new city, 300 *miles north of Ispahan,* 4 or 5 miles in circumference, and *in winter, contains* 60,000 *inhabitants*, but during the summer season, the principal part of the inhabitants remove, on account of the unhealthfulness of the climate.

3. *Shiraz, 5th class*, 160 *miles south of Ispahan*, is beautifully *situated*, and is *distinguished* for its colleges, and also for its wine, which is esteemed the best in Asia.

Q. What is the prevailing *religion?*

A. Mahometanism.

Q. What is the *character* of the Persians?

A. They are gay, polished, and hospitable ; but treacherous, avaricious, and wanting in industry and enterprise.

The *mountainous parts* of Persia, and also the *deserts*, are *inhabited by* the Iliats, or wandering shepherds, who often plunder the inhabitants of the cultivated districts, and have rendered some regions nearly desolate.

Q. What can you say of the Persian *language?*

A. It is the most celebrated of all the oriental tongues for strength, beauty, and melody.

No *printing* is allowed in Persia ; *learning* is of course at a low ebb. The number of people employed on the manuscripts of Persia is almost incredible. The learned *profession in the greatest esteem* is that of medicine ; but every dose must be administered in some lucky hour, fixed by an astrologer.

Q. Which are the principal *productions?*

A. Wine, silk, corn, rice, fruits, and drugs.

Pearls of great value *are found in* the Persian gulf, and men are constantly employed in diving for the oysters which contain them. The *best pearls are found in* water from 12 to 20 fathoms deep. The *divers* seldom live to a great age. Their bodies break out in sores, and their eyes become weak and blood shot. It is said that they can *remain under water* five minutes.

Q. What is the form of *government?*

A. An absolute monarchy.

Questions—Map of Asia.

How is Persia, including E. Persia bounded N. E. S. W.? What desert does P. Proper contain? What sea lies on the N.? What gulf in the S.? Which is the chief town, and which the capital? *How are they situated? Where is Shiraz?*

EAST PERSIA ; OR CABUL, AND BELOOCHISTAN.

Q. How is East Persia *divided?*

A. Into Cabul, which occupies the northern part, and Beloochistan, which lies in the southern part.

1. The *surface, soil,* and *climate* of East Persia, is much

varied. Extensive, barren, parched wastes abound. Portions of the country are well watered and productive; the valley of *Cashmere* is celebrated for its beauty and fertility.

2. In the *valleys*, which are well watered, the *soil* is fertile; but there are extensive wastes. The *desert* of Beloochistan, is 300 miles *long*, and 200 *broad*.

Q. Which is the principal *river* of E. Persia?

A. The Indus.

The Indus, *2d class*, is supposed to *rise* on the western side of the Himmaleh Mts.—It *flows* west of south, and *falls* into the Sea of Arabia. It is *navigable* for vessels of 200 tons, for 850 miles.

Q. Which is the *chief city* and *capital*?

A. C...l.

1. *C...l, 2d class*, is *situated* on the northeast, near a branch of the Indus; it has an extensive trade with Tartary, Persia, and India.

2. The *other principal towns* are Herat, *3d class*, between Cabul and the Caspian, a place of great trade; and Balk, *large town*, north of Herat, a large, populous city, the centre of trade between Independent Tartary, and Hindoostan.

Q. What can you say of the *inhabitants*?

A. They consist of Hindoos, Tartars, Beloochees, Parsees, and Afghans, which last tribe is the most powerful, and its Khan, or chief, is the king of the whole country.

Q. What articles of *manufacture* deserve notice?

A. The shawls of Cashmere, celebrated for their unrivalled beauty.

These shawls are made of the wool, or hair of a kind of goat, found only in Thibet. Three persons are usually employed on the finest of these shawls for more than a year. Sometimes they are not able to complete a quarter of an inch a day.

Q. Which is the *chief town* and *capital* of Beloochistan?

A. K...t.

1. *K...t, 6th class*, is surrounded by a wall, and is the residence of the Khan. The streets are narrow and dirty, and the houses are built of half burnt brick.

2. Very little is *known of the country of Beloochistan*. The inhabitants are of several tribes, the two principal of which are the Beloochees, who much resemble the plundering Arabs; and the Brahooes, who are peaceable, and live chiefly by means of their flocks. The Khan, or chief of

Kelat, is acknowledged as superior to the chiefs of the other tribes.

Questions.—Map of Asia.

How is East Persia bounded? How is it divided? What mountains are on the north? What is the great river of E. Persia? *Where does it rise? Which way does it flow? Where does it empty?* Which the chief town of Cabul? Which of Beroochistan?

Independent Tartary.

Tartars.

Q. How is Independent Tartary *bounded?*

A. N. by R...; E. by the C... E...; S. by P...; W. by P..., the C... S..., and R...

1. The *face of the country* is considerably diversified, but the northern part is more level than the southern.

2. The *climate* is warm and pleasant, and the *soil* in many parts good. Agriculture, however, is much neglected.

3. *Square miles*, supposed to be 1,000,000; *Pop.* 3,500,000.

Q. Which are the principal *mountains?*

A. The chain of Belur Tag.

This chain is on the east of Independent Tartary, and *separates* it from the Chinese empire.

Q. Which are the principal *rivers?*

A. The Oxus and Sihon.

1. The *Oxus*, 3d class, *rises* in the S. E., and *flowing* northwest, *falls* into the Sea of Aral.

2. The *Sihon*, 3d class, *rises* in the mountains of Belur Tag, and *flowing* northwest, *falls* into the Sea of Aral.

Q. Which are the principal *seas?*

A. The Caspian, and Sea of Aral.

Q. Which is the *chief town ?*

A. S...d.

1. *S...d, 3d class, is on* a branch of the Oxus; it was formerly a seat of science, but retains little of its former splendor.

2. *Bucha'ria, 3d class,* is 100 *miles west of Samarcand,* it contains several colleges for instruction in the Mahomedan law.

Q. What is the prevailing *religion ?*

A. Mahometanism.

Q. What is the *form of government ?*

A. Monarchical.

The inhabitants are separated into independent tribes which wander with their flocks according to their pleasure or convenience, each tribe being governed by a Khan, or chief.

Questions.—Map of Asia.

How is Independent Tartary bounded N. E. S. W. ? What sea does it contain ? What are its rivers ? *Describe them.* What mountains lie on the east ? What sea on the west ? Which is the chief town ? *How situated ?*

Hindoostan.

View of Calcutta.

Q. How is Hindoostan' (or Hindoos'tan) *bounded?*

A. N. by T..., from which it is separated by the ᴜ... mts.; E. by the ʙ... ᴇ..., and ʙ... of ʙ...; S. by the ʙ... of ʙ..., and ɪ... ᴏ...; W. by s... of ᴀ... and ᴇ.. ᴘ...

1 *Hindoostan is often called* India within the Ganges. or

India on this side the Ganges. More than half the country has either been conquered by Great Britain, or the governments are tributary to her.

2. The following table shows the estimated extent and population of Hindoostan:

	Geog. square miles.	Population.
British Hindoostan,	415,000	53,500,000
British Allies and Tributaries,	169,000	17,500,000
Independent states,	495,000	30,000,000
Total of Hindoostan,	1,079,000	101,000,000.

3. *Northern Hindoostan* is mountainous. On the sea coast it is level, but hilly in the interior.

4. *The climate of the southern part* is generally hot and unhealthful; the climate of the northern part is temperate and pleasant.

5. The *soil* is in general very fertile, producing two harvests in a year; the former in September and October; the latter in March and April.

Q. Which are the principal *mountains?*

A. The Himmaleh on the north, and the Gauts on the Malabar coast.

The *Himmaleh* mountains, 1st class, are the most lofty and rugged of all the mountains on the globe. There are 21 *peaks estimated to be more than* 20,000 *feet high;* and the *highest called* Dawalage'ri is 27,677 feet above the level of the sea. The Gauts are of the 6th class.

Q. Which are the principal *rivers?*

A. The Ganges, and Burrampoot'er.

1. The *Ganges*, 2d class, *rises* in the Himmaleh mountains, and *flowing* southeast *falls* into the bay of Bengal. It is *navigable* 1350 miles. At 800 miles from the sea, it is at its lowest state, 30 *feet deep.* Thirty thousand *boatmen are employed upon it.* The Hindoos esteem its waters sacred.

2. The *Burrampooter*, 2d class, *rises* in Thibet, not far from the source of the Ganges, *flows* first easterly, and then south of west. It *falls* into the Ganges, 40 miles above its mouth.

Q. What *island* lies near the coast of Hindoostan?

A. The island of Ceylon, which belongs to Great Britain, and is distinguished for its *cinnamon,* which is esteemed the best in the world.

1. The *soil* of Ceylon is very fertile and the climate healthful. Besides cinnamon, it *produces* ginger, pepper, sugar, and cotton.

2. Columbo, the *capital*, is of the *4th class*. The population of the whole island is estimated at 1,500,000. The religion is idolatry; but missionaries from England and America are labouring with success, in several parts of the islands.

Q. Which is the *chief town* and *capital* of Hindoostan?

A. C...a.

1. *C...a, 1st class, is on* the Hoogly, an outlet of the Ganges, about 100 *miles from the sea*. It is one of the largest cities in the world. It is a place of immense *commerce* in sugars, silks, muslins, and calicoes. The houses of the English, are brick; those of the natives, mostly mud cottages.

2. *Benares, 1st class*, the *famous seat* of Brahminical learning, is regarded by the natives as a holy city. It *lies* on the Ganges. It is a place of great wealth and trade.

3. *Bombay, 2d class*, is *situated* on an island, 10 miles in length, near the west coast, and commands the whole trade of the northwest coast of the country.

4. *Serampore, large town*, 12 miles north of Calcutta, is the head quarters of the Baptist missionaries. *Jug'gernauth*, situated on the eastern coast, is a celebrated place of Hindoo worship. The number of pilgrims who annually visit this place, is estimated at 1,000,000, most of whom never return.

5. Hindoostan contains many *other cities* of great population; Surat, *1st class*, Madras, and Lucknow, *2d class*, and Delhi, *3d class*.

Q. Who are the *inhabitants* of Hindoostan?

A. The great mass are Hindoos, who are tame, timid, superstitious, selfish, and vicious.

The *Hindoos are divided* into classes called *castes*: 1. the *Brahmins*, or priests; 2. the *soldiers*; 3. the *merchants* and *agriculturists*; 4. the *soodas* or labourers. These castes are kept entirely distinct, and are never allowed to intermarry, nor even to eat and drink together.

Q. What is the *religion* of the Hindoos?

A. Idolatry.

1. The Hindoos have an immense number of temples, which are *called* pagodas; their *sacred books* are called Vedas. They have an *ancient commentary* on the Vedas, called the Shaster. *See page* 19.

Q. Which is the principal article of *manufacture*?

A. Cotton; next to which are the articles of silk, wool, leather, and salt petre.

Q. Which are the principal *productions*?

A. Rice is most extensively cultivated; but attention is also paid to the growth of cotton, wheat, indigo, sugar, and tobacco.

Among the *beautiful natural productions* of Hindoostan is the Banian Tree; the branches of which strike down, and take root from stems or trunks, so that each tree becomes a grove. The *most celebrated* of these trees is on an island in the river Nerbuddah. It has more than 3000 trunks, measuring about 2000 feet in circumference, and 7000 persons may repose under its shade.

Q. Under whose direction is the *commerce* of Hindoostan ?

A. Under that of Great Britain ; whose merchants export vast quantities of cotton, piece goods, rice, salt petre, indigo, silk, and sugar.

Q. What is the *government* of Hindoostan ?

A. The English part of Hindoostan is under the direction of a governor general, appointed by the king of England ; the governments of the independent states are absolute monarchies.

Questions.—Map of Asia.

How is Hindoostan bounded? What mountains separate it from Thibet ? What mountains are on the Malabar coast ? Which is the great river of Hindoostan ? *Describe it; also describe the Burrampooter.* What are the eastern and western coasts called? What is the southern point of Hindoostan called ? Which is the chief town and capital ? *What is its latitude ? Where is Bombay ? Where Madras ? Delhi ? Juggernauth ?* What island lies on the southern coast of Hindoostan ? What is the chief town and capital of the island ? What groups lie westerly ?

Farther India, or India beyond the Ganges.

Elephant.

Q. How is India beyond the Ganges *bounded?*

A. N. by the c... e... ; E. by the c... s...; S. by

the c... s..., and B... of B...; W. by the B... of B...,
and H...

Q. What countries does India beyond the Ganges *include?*

A. The Birman empire, Malacca, Siam, Laos,
Cambodia, Chiampa, Cochin China, Tonquin, and
Assam.

BIRMAN EMPIRE.

Q. How is the Birman Empire *situated?*

A. In the west part of Farther India, bordering
upon Hindoostan and the Bay of Bengal.

1. It *includes* the ancient kingdoms of Ava, Pegu, Arracan,
and Cassay, or Meckley.

2. The *country in the northern part is mountainous,* but the
southern part consists chiefly of extensive valleys and plains.

3. The *climate* is subject to great extremes of cold and
heat, but in general is healthful.

4. The *soil* is fertile, yielding great quantities of rice.

5. The *population* is supposed to be 17,000,000, and of all
India beyond the Ganges from 30 to 40,000,000.

Q. Which is the principal *river?*

A. The Irrawad'dy.

The Irrawad'dy, 2d *class,* is supposed to *rise* in the eastern
part of Thibet; it *flows* southerly, and *falls* into the Bay of
Bengal.

Q. Which is the *chief town* and *capital?*

A. U...a.

1. U...a, 3d *class, is situated* on the Irrawaddy, 400 miles
above its mouth.

2. The *other principal town* is *Rangoon,* 4th *class,* on a
branch of the Irrawaddy, 30 miles from the sea, and is the
great seaport of the Empire.

Q. What is the *religion* of the Birmans?

A. They worship the *Buddhu,* but the image
which represents him is called *Godama.*

Buddhu is represented as a young man with a placid countenance, and usually sitting cross-legged on a throne. Some
of the Godamas or images are of gigantic magnitude.

Q. What is the *character* of the Birmans?

A. They are more lively, intelligent, and enter-

prising, than the Hindoos, but are impatient, irascible, and cruel: they are fond of poetry and music.

Q. What is the *government?*

A. An absolute despotism.

MALACCA, SIAM, LAOS, CAMBODIA, CHIAMPA, COCHIN CHINA, TONQUIN, AND ASSAM.

Q. How are these countries *situated?*

A. South and east of the Birman empire, and
south of China. Malacca is a long peninsula, form-
ing the most southern part of Asia.

1. The *country on the rivers* is level, and is sometimes
overflowed. At a distance from the rivers the country rises
into hills and mountains.

2. *Population* of Laos 3,000,000; of Siam and Malacca
2,000,000; Assam 2,000,000; and the remaining countries
from 6 to 16,000,000.

Q. Which are the principal *rivers?*

A. The Meinam (Mee-nam) and Cambodia, or
Japanese river.

1. The *Meinam, rises* in the mountains of Thibet, and
running through the Birman empire and Siam, *flows* southerly into the gulf of Siam.

2. The *Cambodia*, 1st class, or Japanese, *rises* in the
mountains of Thibet, and *running* through the S. W. part of
China, and through Laos and Cambodia, *flows* southerly into
the China sea.

Q. Which are the *capitals?*

A. *M...a* is the capital of Malacca. *S...m* is the
capital of Siam. *C...a* is the capital of Cambodia.
K...o or *Cachao*, is the capital of Tonquin, and *T...e*
the capital of Cochin China.

M...a belongs to the 6th *class; S...m* to the 3d *class; C...a*
is a *small town; K...o*, 5th *class; T...e*, uncertain.

Q. What is the *religion* of the inhabitants?

A. Paganism.

Q. What is their *character?*

A. The *Malays* are ferocious, restless, treacherous, and piratical; the *Siamese* resemble the Birmese,
but have made greater advances in civilization and
science; the *Cochin Chinese* are open, familiar, gay,
talkative, and courteous.

Q. What *islands* lie in the bay of Bengal?

A. The Andaman' and Nicobar' islands.

These *islands produce* great numbers of birds' nests made of a viscous substance, resembling isinglass. These, when dissolved in broth, become a jelly, which are highly esteemed by the Chinese.

Q. What is the *government* of these countries?

A. Absolute monarchies.

Questions.—*Map of Asia.*

How is India beyond the Ganges bounded? What empire lies in the western part? Which are some of the other countries? Which are the principal rivers? *Describe them.* Point out the capitals of the several countries. What is the southern part of India beyond the Ganges called? What island lies south? What groups lie west of the peninsula of Malacca? Where is the gulf of Siam? Of Tonquin?

Chinese Empire.

View of Pekin.

Q. How is the Chinese Empire *bounded?*

A. N. by R...; E. by the P... O...; S. by the C... s.., India, and H...; W. by P... and T...

Q. What *countries* does the Chinese Empire include?

A. China Proper, and its Tributary States.

CHINA PROPER.

Q. How is China *bounded?*

A N. by C... T...; E. by the P... O..; S. by the c... s..., and India; W. by India and T...

1 The *face of the country* is agreeably diversified by

16*

mountains and hills, and is intersected in every direction by rivers and canals.

2. The *soil* of China is in general very fertile.

3. The *climate* of the north is said to be cold in winter. The heat of the southern part is great in summer. Yet people often attain to a great age.

4. The *square miles* of China Proper are estimated by some to be 1,300,000; by others, 2,000,000. The *inhabitants*, from 160,000,000 to 333,000,000.

Q. What *seas* are contiguous to China?

A. The Yellow and China Seas.

Q. Which are the principal *rivers?*

A. The Hoang-Ho, and Kiang-Ku.

Both of these rivers, 1st *class*, *rise* in the mountains of *Thibet*, and both *flow* into the east of China; the *former winding* to the north, the *latter* to the south.

Q. Which is the *chief town* and *capital?*

A. P...n.

1. *P...n*, 1st *class*, is *situated* in the northeast of China. It is 14 *miles in circumference*, and is surrounded by a wall 30 *feet high*, with 9 *gates*. The streets are straight and wide. The *houses* are only one story *high*

2. *Nankin*, 1st *class*, was *formerly* the imperial city, but has greatly fallen from its ancient splendour. It is the first city in China in *regard to the manufacture of silks*, crapes, and nankeens. It has a *famous* porcelain *tower*, 200 feet high, and 40 in diameter. It is composed of 9 *stories*, and is ascended by 884 *steps*.

3. *Canton*, 1st *class*, is the only port to which European and American vessels are admitted. The temples, magnificent palaces, and courts, are numerous. 300,000 persons reside in barks, which form a kind of floating city. They touch one another, and are so arranged as to form streets.

4. China contains many other cities, each of which is said to have more than a million of inhabitants.

Q. What is the *religion* of the Chinese?

A. The religion of the Chinese has remained the same for ages; it does not agree with that of any other country. The people believe in a supreme being called Fo, to whom is attributed universal knowledge, power, and perfect justice. Temples, full of images, erected to Fo, abound in China. This religion is not established by government.

Q. What can you say of the *roads* and *canals* of China?

A. They surpass all others in the world.

The great canal between Pekin and the Kiang-Ku is 600 miles long. It is said to have employed 30,000 men upwards of 40 years in its construction.

Q. What *great curiosity* does China contain?

A. A wall, which extends 1500 miles along the north of China, separating it from Tartary.

This wall passes over high mountains, wide rivers, supported by arches, and across deep valleys. It is 30 feet *high* on the plain, and 15 or 20 feet when carried over rocks and elevated grounds. It is so *thick*, that six horsemen can easily ride abreast upon it. It is said to have been built more than 2000 years ago, to prevent the inroads of the Mogul Tartars into China.

Q. What is the state of *education* in China?

A. Public schools are numerous: the children of the poor, however, are neglected; but in the higher classes considerable attention is paid to learning.

The Chinese language is very peculiar. They write in a kind of hieroglyphical mode. *The number of the characters used*, is from 35,000 to 40,000, one third of which is more than sufficient for the common purposes of life.

Q. What is the *character* of the Chinese?

A. They are mild, courteous, affable, and ceremonious; at the same time vain, timid, artful, treacherous, and jealous of strangers.

1. The Chinese are allowed to have several *wives*, but the women are in a state of abject degradation. They are sold in marriage, without their own choice. They are *esteemed most beautiful when* very corpulent, and have small feet. Their feet are confined in infancy to prevent their growing.

2. The men and women *dress* much alike. The higher classes of both sexes sometimes suffer their nails to grow to an immoderate length. At night they are carefully enclosed in cases to prevent their being broken. They are always preserved clear and transparent.

3. *Dress is regulated* by law. White is worn for *mourning*. Children treat their parents with great reverence. To strike a parent is punished with death. Parents who are unable to support their female children, are permitted to throw them into the river.

Q. What is the *government* of China ?

A. An absolute monarchy ; but the government is generally administered with much kindness.

Q. What is the state of *agriculture ?*

A. It is carried to the highest perfection.

At the *vernal equinox* the *emperor himself,* and all the great officers of the empire, *perform the ceremony of* holding the plough.

Q. Which is the most celebrated *vegetable production* of China ?

A. The tea plant.

1. *Tea is the leaf of* a shrub which grows to the *height* of from 8 to 10 feet, and yields crops of leaves 3 years after being sown, but requires to be renewed every 5 or 6 years.

2. China *also produces* the camphor tree, from the roots of which camphor is obtained, by distillation. Also the tallow tree, paper mulberry tree, &c.

Q. Which are the principal *exports?*

A. Tea, silk, nankeens, porcelain, sugar, cinnamon, and camphor

Q. What *islands* belong to China ?

A. The islands Formosa and Hainan, and the Leeoo-Keeoo, (Loo-Koo) islands.

Questions.—Map of Asia.

How is the Chinese Empire bounded N. E. S. W. ? Which are the great divisions of the Empire ? How is China bounded N. E. S. W.? Where is the Chinese wall ? Which are the two principal rivers? *Describe them. What great canal do you notice ? What city and river does it connect?* Which is the chief town and capital ? *How situated? What is its latitude and longitude? What distinguished city in the U. States is in about the same latitude? Which way from Pekin is Nankin? Which way is Canton?* What island lies near the coast ? Where are the Leeoo-Keeoo islands which belong to China ?

TRIBUTARY STATES OF CHINA.

Q. What countries are *tributary* to China ?

A. Corea, Chinese Tartary, and Thibet.

COREA.

Q. Where is Corea *situated?*

A. It is a small peninsula northeast of China, but is little known.

The *southern* part of Corea is fertile and populous. King-kiteo (King-ke-tow') is the *capital.* The inhabitants are idolaters.

Questions.—Map of Asia.

How is Corea situated ? Which is the chief town and capital ?

CHINESE TARTARY.

Q. How is Chinese Tartary *bounded ?*

A. N. by R...; E. by the s... of o..., and s... of J...; S. by c..., c..., and t...; W. by p..., and i... t...

1. The *greater part* of Chinese Tartary is a *sandy desert*, barren and destitute of water, elevated about 8 or 9000 feet above the level of the sea, and encircled by the mountainous ranges of Himmaleh, Altay, Belur Tag, and China.

2. The *climate*, from the elevation of the country, is cold; the *soil* is unproductive.

3. The *extent* and *population* are very uncertain.

Q. Which is the principal *river ?*

A. The Amour, or Saghali'en.

The *Amour, 1st class*, *rises* in the Altay mountains, and *flowing* east, *enters* the sea of Ochotsk.

Q. What can you say of the *inhabitants ?*

A. Very little is known about them; they consist of various tribes, and lead a wandering and pastoral life.

Questions.—Map of Asia.

How is Chinese Tartary bounded N. E. S. W. ? How is it separated from China ? Which is the great river ? *Describe it. What desert in the interior ?*

THIBET

Q. How is Thibet *bounded ?*

A. N. by c... t...; E. by c...; S. by the b... e... and h...; W. by h...

1. The *face of the country, climate*, and *soil*, very much resemble those of Chinese Tartary, excepting that it is more mountainous, and the valleys more fertile.

2. *Square miles*, about 360,000. *Pop.* supposed to be about 16,000,000.

Q. Which is the *chief town* and *capital ?*

A. L...a.

L...a is 500 *miles northeast of* Calcutta, and is *celebrated as* the residence of the Grand Lama.

Q. What is the character of the *Thibetians ?*

A. They are ignorant, mild, indolent, timid, and superstitious.

Q. Which is the *religion* of the Thibetians?

A. They worship the governor, or **Grand La**r who administers the government, under the empe of China.

The *Grand Lama inhabits* a magnificent temple, and served with the most profound reverence, by a **great** numb of priests. He is seldom *seen*. *When he dies, the prie pretend* that his soul enters the body of some **infant**, who they select, and with great pomp establish in his plac Hence, his worshippers pretend that he is immortal.

Questions.—Map of Asia.

How is Thibet bounded N. E. S. W.? What large rivers rise the Mts. of Thibet? Which is the chief town and capital? *Ho situated?*

Japan.

Sedan Chair.

Q. Of what is the empire of Japan *composed?*

A. Of a cluster of islands, lying east of Chinese Tartary.

1. Of these islands, Niphon is the *largest*, and is said to be 700 miles *long*, and on an average 80 *broad*. Several of the other islands are of considerable size.

2. The *face of the country* is equally diversified by moun tains, hills, and valleys, and is well watered. Some of the mountains are said to be very high.

3. The *soil*, not naturally the most fertile, has been highly improved by the industry of the Japanese.

4. The *extent* of the three largest islands is variously esti mated from 90,000 to 180,000 *square miles;* the *population* from 15 to 50,000,000

Which is the *chief town* and *capital?*

A. J...o, on the northeast side of Niphon.

1. J...o, *1st class*, is said to be one of the most splendid cities in the world. It is *situated* in a bay, on the southeast side of the island of Niphon.

2. Meaco, *1st class*, is the second city, *situated* 160 miles S. W. of Jeddo, and is the store house of all the manufactures of the empire.

3. *Nangusaki, large town*, on the island of Ximo, is the only port visited by foreigners.

Q. What is the *religion* of the Japanese?

A. Paganism.

Q. What is the *character* of the Japanese?

A. They are intelligent, ingenious, kind, and friendly, with a high sense of honour; and the obligations of friendship.

The *customs of this people* vary greatly from ours. We uncover our heads out of respect; but they uncover their feet; we rise up to receive a visiter, they sit down; are fond of white teeth, they of black; their *common* drinks are made quite hot, we drink ours cold; we mount our horses on the left side, they on the right side.

Q. What can you say of *learning* among them?

A. They are said to be in general well informed, and to have several respectable colleges, or public schools, among them.

Q. What *minerals* are found in Japan?

A. It is said to be the richest country in the world for gold; silver, copper, and lead, also abound.

Q. What is the principal *production?*

A. Rice; millet, wheat, and barley, are also cultivated.

The *teas* of Japan of all sorts are much finer and better cured, than those of China.

Q. What can you say of their *manufactures?*

A. They excel in the manufacture of silk, cotton porcelain, and *japanned* ware.

Q. With *whom* do they *trade?*

A. With the Dutch and Chinese only.

The *Japanese formerly traded with* many countries, but in consequence of a suspicion that some missionaries, who had been introduced by the Portuguese, were attempting to subvert the empire, *all intercourse with foreign countries* prokibited, excepting with those already mentioned.

Q. What is the *government?*

A. An absolute, hereditary monarchy.

Questions.—*Map of Asia.*

How are the islands of Japan situated? Which are the princip of them? Which is the largest? What separates them from Core: On what island is the chief town and capital? What is the name the place? *Where is Meaco? Where Nangasaki, the chief pla of trade?*

AFRICA.

Egyptians.

Q. How is Africa *bounded?*

A. N. by the M... s... which separates it from E...; E. by the R... s... which separates it from A..., and the I... o...; S. it terminates in the s... o... at Cape Horn; W. by the A... o...

1. *Africa* is one of the divisions of the globe, and the third in extent. Comparatively little is known of it, excepting the countries lying on the coast.

2. The *climate* of Africa is hot, and to foreigners quite unhealthful.

3. The *face of the country* is in general very level, especially the interior, which consists of immense sandy and barren deserts, with scarcely a verdant spot, and with but few springs of water.

4. The *soil* of the well watered parts of Africa is fertile; the deserts are nearly a barren waste.

5. The *length* of Africa is nearly 5000 miles, and its greatest breadth 4500. Square miles, 11,000,000 souls say

13,000,000. Its *population* is variously estimated from 30 millions to 150 millions.

Q. Which are the principal *mountains?*

A. The two principal mountains are the Atlas mountains, and the Mountains of the Moon.

1. The *Atlas chain commences* north of cape Baja'dor, and *runs* northeast along the coast to Cape Bon. This chain *divides* the greater part of Barbary from the vast Desert of Sahara. The mountains which form the eastern boundary of Morocco are the *loftiest. Highest peaks*, 4th class.

2. The *mountains of the Moon commence* near cape Verde, and *run* east, almost to cape Guardafui, (Gwor-da-fwee') a distance of near 3000 *miles*. The *eastern part* of the chain is also *called* the Abyssinian Alps, and the *western*, the Mountains of Kong. The *highest peaks*, 4th class.

Q. Which are the principal *seas* and *straits?*

A. The Mediterranean Sea, and Strait of Gibraltar, which separate Africa and Europe; and the Red Sea and Strait of Babelmandel, which divide it from Asia.

Q. Which are the principal *rivers?*

A. The Nile, Ni'ger, Con'go, Gam'bia, Senegal (Sen-e-gawl') and Rio Grande, (Re'o-Grand'.)

1. The *Nile, 1st class*, is one of the most celebrated rivers on the globe. It is *formed* of two principal branches, the *largest* of which *rises* in the Mountains of the Moon, and the *other* in the mountainous parts of Abyssinia. Its *course* is notherly, and it *flows* into the Mediterranean.

2. The *Niger, 1st class*, probably, *rises* in the western parts of Africa, and *flows* a great distance eastward, passing Sego and near Tombuctoo. Its *termination* is unknown.

3. The *Congo* or *Zaire*, and *Senegal, 2d class*, the *Gambia, 3d class*, and the *Rio Grande, 5th class*, all *rise* in the western part of the Mountains of the Moon, and *flow* west into the Atlantic.

Q. Which are the principal *capes?*

A. Cape Bon in the north; Cape Guardafui in the east; Cape of Good Hope in the south; and Cape Verde in the west.

Q. Which are the principal *lakes* in Africa?

A. Tchad, (Chad) Dembea, and Maravi.

17

Q. What great *desert* does Africa contain ?

A. The Saha'ra desert, which is by far the great
est desert in the world.

This desert *extends* from the Atlantic on the west, to the
Nile on the east, 3000 *miles ;* and from the Barbary states on
the north, from 800 to 1000 miles south. Excepting a few fer
tile spots, which serve as resting and watering places for the
caravans in their journies, *this desert presents the spectacle*
of a naked, burning plain of sand. Sometimes caravans
die from thirst. In 1805, a caravan of 2000 men, and 1800
camels, entirely perished, from this cause.

Q. What *articles of commerce* are most extensively obtained
in Africa ?

A. Gold and ivory, but above all, slaves.

Q. What *wild animals* are found in Africa ?

A. Lions, elephants, tigers, and panthers. The
rivers abound with crocodiles. Of tame animals the
camel is the chief.

Q. Which are the principal *productions ?*

A. Sugar, salt, gold dust, ivory, various kinds of
fruit, drugs, gums, pearls, and the common neces-
saries of life.

Q. What can you say of the *inhabitants ?*

A. They consist chiefly of Moors and Negroes;
the former of whom are intolerant, treacherous, and
sanguinary; the negroes are mild and amiable, but
wanting in energy and enterprise.

Q. How may Africa be *divided ?*

A. Into Northern, Western, Southern, Eastern,
and Central Africa; together with the African
Islands.

Questions.—Map of Africa.

In what direction does Africa lie from the other divisions of the
globe ? How is it bounded N. E. S. W. ? What isthmus unites it
to Asia ? What is the most northern cape of Africa ? Which the
most southern ? Which are the eastern and western capes ? *Between*
what latitudes and longitudes does Africa lie ? What other capes
do you notice round the coast of Africa ? What sea lies on the east ?
What on the north ? What strait enters the Red sea ? What channel
on the eastern coast ? *How wide is this channel ? Ans.* 270 *miles.*
What bays on the eastern and western coasts ? What gulf lies near
the equator ? What mountains run through the centre of Africa ?

What is the western part of these mountains called? What mountains in the northern part? *What mountains give name to the Atlantic Ocean? Ans. Atlas.* What chain of mountains in S. Africa? What in east? What great desert lies in the northern part of Africa? *What is the extent of this desert? Ans. About 3000 miles from east to west, and 1000 from north to south.* What other deserts does Africa contain? *Ans. Desert of Barca, near Barca; the desert of Lybia south of Barca; the desert of Nubia in Nubia; and the desert of Cimbelas south of Benguela.* Where is lake Maravi? Where Dembea? Which are the two largest rivers in Africa? *Where do they rise, and which way do they flow?* Where are the Senegal, Gambia, and Rio Grande? *In what mountains do these rivers rise, which way do they flow, and where do they empty?* Where are the rivers Zaire, Orange, and Cuama? What large island lies east of Mozambique? What islands east of Madagascar? What north? Where is St. Helena? Ascension? Where the Cape Verde islands? Canary islands? Madeira? The Azores? Which are the five great divisions of Africa.

I. Northern Africa.

Q. What countries does Northern Africa *comprehend?*

A. E'gypt, and the Bar'bary states.

I. EGYPT.

Grand Cairo.

Q. How is Egypt *bounded?*

A. N. by the M... S...; E. by the I... of S... and the R... S...; S. by N...; W. by the great D... and Barca.

1. Egypt is *divided* into Upper and Lower Egypt. *Upper Egypt extends* from the city of Syene to Cairo; *Lower Egypt* from Cairo to the Mediterranean, and is *styled* the Delta

2. The *cultivated* part of Egypt is a vale 15 or 20 miles broad, through which the Nile flows. Beyond the valley the country is barren and mountainous.

3. The *climate in the summer* is very hot, and the dreadful ravages of the plague are frequently experienced. The winters are mild and agreeable. Rain is almost entirely unknown. For about 50 days in the spring, Egypt is *liable to* the Simoon, but it rarely lasts more than three days.

4. The *soil* of the vale, watered by the Nile, is very fertile.

5. *Length*, 600 miles ; *breadth*, from 2 to 300 ; *square miles*, 120,000. *Pop.* 2,500,000.

Q. Which is the principal *river* of Egypt ?

A. The Nile.

The *Nile*, 1st *class*, is *formed* of two branches, *one* of which *rises* in the Mountains of the Moon ; the *other* in the mountains of Abyssinia. It *flows* northerly and *falls* into the Mediterranean. It annually overflows its banks, and spreads over the country like a sea. When the waters subside, they leave a mud or slime, which greatly enriches the soil. The *rise of the waters commences* about the middle of June, and increases about 10 weeks. The *cause of this rise* is the rains, which fall in Abyssinia, from June to September.

Q. Which is the *chief town* and *capital*?

A. G...d C...o.

1. *C...o*, 2d *class*, is *situated* near the east bank of the Nile, with which it is connected by the canal of Kalische.— It *carries on an extensive commerce*, by means of caravans, with Syria, Arabia, and the interior of Africa. One of the *greatest curiosities* is the well in the castle, called *Joseph's Well*, which is sunk 276 feet in a solid, but soft rock. It has a winding stair case descending to the bottom.

2. *Alexan'dria*, 6th *class*, 125 miles N. W. of Cairo, was once a splendid city, containing according to some, 700,000 inhabitants. It was *celebrated* for its learning, commerce, and magnificence. Among the *remains* of its former grandeur, are Pompey's Pillar, Cleopatra's Needle, the Cisterns, and Catacombs. It formerly *contained a library*, consisting, according to some, of 700,000 manuscript volumes, but which was burnt in the year 638.

3. The *other principal towns* are Damiet'ta, 5th *class*, on the eastern branch of the Nile, the great emporium of commerce between Egypt and Syria. Roset'ta, 6th *class*, on the western branch of the Nile, a place of considerable commerce ; and Suez, *small town*, at the northern extremity of the Red sea, 59 miles E. of Cairo, the refreshing place of caravans, passing from Egypt into Syria and Arabia.

Q. What can you say of the *inhabitants ?*

A. They consist of Copts, who are represented as acute, but sober, avaricious, and grovelling; and Arabs, Turks, and Jews.

The Copts are the *original inhabitants of Egypt*, and *amount* to about 200,000. The Arabs constitute two thirds *of the population.*

Q. What can you say of the *religion* of the inhabitants ?

A. The Copts profess christianity ; all other classes, except the Jews, are Mahometans.

Q. Which are some of the remarkable *remains of antiquity ?*

A. Pyramids, statues, temples, and catacombs.

1. The *Pyramids* were anciently accounted one of the seven wonders of the world, and continue to be objects of admiration, as monuments of ancient art and power. *They are* square piles of stone rising to a point. Three large ones stand opposite to Cairo. The *largest covers* more than 11 acres of ground, and is 499 feet high. They *were probably intended as* sepulchres for the kings of Egypt. But *when they were erected,* history does not inform us.

2. *Near one of these pyramids is* the celebrated Sphinx, or a statue of a huge monster, cut in the solid rock, having the face of a virgin, and the body of a lion.

3. *The Catacombs are* galleries under ground, where dead bodies were anciently deposited. Mummies, or bodies embalmed, *are taken from these catacombs,* which were deposited there more than 3000 years ago.

Q. What is the *government* of Egypt ?

A. Egypt has long been subject to Turkey, and governed by a Pacha ; who has lately set up an independent monarchical government.

Questions.—Map of Africa.

How is Egypt bounded N. E. S. W.? Which is the chief town and capital ? *How situated?* Which is the great river of E. ?— *Which way does it flow, and where does it empty ?*

17*

II. BARBARY STATES.

View of Algiers.

Q. How are the Barbary States *situated?*

A. They occupy that long, narrow country, lying along the м... s... on the N. ; and having on the E. the м... s..., and E... ; S. the s... D...; and W. the A...

1. The *space between the Atlas range and the sea,* is from 50 to 200 miles wide, and is mostly a level and well watered country. The *country south* of the mountain is generally barren.

2. The *climate* is temperate and healthful, excepting that it is sometimes subject to the plague.

3. The *soil* of the level parts is fertile, but badly cultivated.

4. The *extent* and *population* are estimated as follows :—

	Population.	Square miles.
Morocco, from 5 to	14,000,000,	290,000,
Algiers, about	2,000,000,	90,000,
Tunis, from 1 to	2,000,000,	72,000,
Tripoli, including } Barca,	2,000,000,	210,000.

Q. Which is the principal *range of mountains?*

A. The Atlas range.

The *Atlas chain, highest peaks, 4th class, extends* through the southern part of Morocco, Algiers, and Tunis. The most lofty summits are in the western part.

Q. What is the *character* of the inhabitants?

A. They are extremely rapacious and cruel, and along the coast are addicted to piracy.

1. The *inhabitants* may be *divided* into four classes, Moors, Jews, Arabs, and Brebers.

2. The *Moors* are the ruling people, and form the chief inhabitants of all the cities. They are ignorant, superstitious, indolent, rapacious, and cruel.

3. The *Jews* are the principal merchants, but are greatly oppressed by the Moors.

4. The *Arabs* live in the interior, and lead a wandering life.

5. The *Brebers*, a strong athletic race, are descendants of the original inhabitants, and occupy the mountainous districts.

Q. Which is the prevailing *religion?*

A. All classes profess Mahometanism, except the Jews.

Q. Which are some of the most remarkable *animals, serpents,* and *insects?*

A. The lion, leopard, and panther; the boa, or serpent of the desert; and the locust.

1. The *boa*, or serpent of the desert, is often 80 feet *long*, and as *thick* as a man's body. It moves so *swiftly*, that no animal can escape it. It will *twist itself around* oxen, and other large animals, crush them and break all their bones; after which it swallows them, and then lies quietly for several days, till the food is digested. In the torpid state it may be killed without danger.

2. The *locust* is a plague of a very destructive nature, and often comes from the deserts in such numbers as to darken the sky, and to devour every green substance.

Q. Which are some of the principal *productions?*

A. Maize, millet, peas, beans, and dates.

Q. What is the *government* of these states?

A. Absolute, despotic monarchies.

Q. What are the *names* of these states?

A. Moroc'co, Algiers', Tu'nis, and Trip'oli.

1. MOROCCO.

Q. How is Morocco *bounded?*

A. N. by the A..., strait of G... and M... s...; E by A...; S. by the s... D...; W. by the A...

Q. Which is the *chief town* and *capital?*

A. F...z is the chief town, M...o the capital.

1. F...z, 3d *class*, the capital of the ancient kingdom of Fez, is 200 miles northeast of Morocco. It is the most splendid city in the Barbary states.

2. *M...o, 5th class, is situated* at the foot of Mount Atlas, 120 *miles from the sea.* It is surrounded by a wall, and is said to have once contained 70,000 inhabitants.

3. *Mequinez, 3d class,* is 35 miles southwest of Fez.

4. *Mogadore', 6th class,* is a noted seaport, 80 miles S. W. of Morocco.

Q. Which is the principal article of *manufacture ?*

A. Morocco leather.

2. ALGIERS.

Q. How is Algiers *bounded ?*

A. N. by the M... s...; E. by T...; S. by the s... D...; W. by M...

Q. Which is the *chief town* and *capital ?*

A. A...s.

1. *A...s, 2d class, is on* the coast of the Mediterranean, 300 miles west of Tunis. It is built on the side of a hill, the houses rising one above another, in the form of an amphitheatre. It is strongly fortified. The *inhabitants are noted for* their piracies and cruelty.

2. The *other principal town* is Constantina, (Con-stan-tee'-na) *3d class,* 160 miles east of Algiers ; it is a strong city, built on a rock.

Q. What country lies south of Algiers and Tunis ?

A. Biledul'gerid, or the country of dates.

The inhabitants are Brebers and Arabs, who are under a kind of subjection to Algiers and Tunis.

3. TUNIS.

Q. How is Tunis *bounded ?*

A. N. by the M... s...; E. by the M... s... and T... ; S. by the s... D...; W. by A...

Q. Which is the *chief town* and *capital ?*

A. T...s.

1. *T...s, 3d class,* is *situated* on a plain, about 6 miles from the head of the Gulf of Tunis, and has a considerable trade. The streets are narrow and dirty, and the houses mostly of one story.

2. The *ruins of Carthage,* once the rival of Rome, are *situated* about 10 miles northeast of Tunis. It once contained 700,000 inhabitants.

4. TRIPOLI.

Q. How is Tripoli, including Barca, *bounded ?*

A. N. by the M... s...; E. by E...; S. by the s.. D..., and F... ; W. by T...

Q. Which is the *chief town* and *capital?*

A. T...i.

T...i, 6th class, lies on the coast of the Mediterranean, aas ea excellent harbour, and is surrounded by high walls.

Q. What *country* is *subject* to Tripoli?

A. Barca, which lies east of Tripoli.

1. *Barca* is mostly a desert, especially in the part next to Egypt.

2. The *chief town and capital* is Derne, situated near the coast, and is *famous* for the exploits of Gen. Eaton, who took it in 1805.

Questions.—Map of Africa.

How is Barbary bounded N. E. S. W.? What mountains lie S.? Name the states of Barbary and their capitals.

II. Western Africa.

Funeral Procession in Congo.

Q. How is Western Africa *bounded?*

A. N. by the tropic of Cancer and c... A...; E. by c... A...; S. by s... A... and A... o...; W. by the A...

1. The *face of the country* is generally level.

2. The *climate* of Western Africa is very hot, and during the rainy season, is very unhealthful for Europeans; much of the *soil* is fertile.

Q. Which are the principal *rivers* of Western Africa?

A. The Congo, or Zaire, Senegal, Gambia, Mesurado, and Rio Grande.

The *Congo* and *Senegal* are of the *2d class;* the *Gambia,* *3d class;* the *Mesurado, 4th class;* and the *Rio Grande, 5th class.* All these rivers, excepting the Congo, *rise* in the Mountains of the Moon, and *flowing* westerly, *empty* themselves into the Atlantic. The *source* of the Congo is unknown. It *flows* southwesterly, and *falls* into the Atlantic.

Q. What is the *character* of the inhabitants?

A. They are in general, indolent, ignorant, and superstitious; but gentle, simple, hospitable, and affectionate.

The *negroes* of the United States, and other countries, are *descended from* the tribes of Western Africa.

Q. Which are the principal *animal* and *vegetable* productions?

A. Lions, Elephants, and Ostriches; sugar cane, indigo, and rice.

Western Africa *also produces* the Boabab, or calabash tree. It grows only to the height of 12 or 15 feet, but its branches extend horizontally 50 or 60 feet; the trunk is 25 or 30 feet in diameter, and the branches 6 or 8 feet. The negroes frequently hollow out the trunks for burying places, and bodies are thus preserved dry, and will resist putrefaction, as if embalmed.

Q. Which are the principal *divisions* of Western Africa?

A. Senegambia, the coast of Sierra Leone, the coast of Guinea, and the coast of Congo.

1. SENEGAMBIA.

Q. How is Senegambia *bounded?*

A. N. by the D... of s...; E. by c... A...; S. by the coast of Sierra Leone; W. by the A...

Q. Which are the principal *tribes* of Senegambia?

A. The Mandingoes, Foulahs, Feloops, and Jaloffs.

1. *The Mandingoes occupy* all *the countries* on the banks of the Niger, the Senegal, and above all the Gambia. They are the most *numerous* of all the tribes of Western Africa. *They are* cheerful, gentle, inquisitive, credulous, and vain.

2. The *Foulahs* also are widely *diffused.* They *occupy* powerful *kingdoms* on the Niger, Gambia, and lower parts of the Senegal. *They are* industrious, hospitable, humane, mild, and polite. Teemboo, *large town,* is the chief town and capital.

3. The *Feloops inhabit* an extensive country on the southern side of the Gambia. *They are* a wild and unsociable race

4. The *Jaloffs occupy* most of the *country* between the low-

or part of the Gambia and the Senegal. *They are* the handsomest negroes in this part of Africa, and are considerably cultivated.

2. COAST OF SIERRA LEONE.

Q. How is the coast of Sierra Leone *bounded?*

A. N. by s...; E. and S. by the c... of g...; W. by the a...

Q. Which is the principal *settlement* in Sierra Leone?

A. The English Colony of Sierra Leone.

This colony *contains* about 13,000 *inhabitants.* It was *formed* with a view to colonize free negroes, and to promote the civilization of Africa. About 20 missionaries are employed here. Freetown is the *chief town and capital.*

Q. What *American Colony* does this country contain?

A. A colony at Cape Mesurado, about 300 miles southeast of Sierra Leone.

This *colony was formed* in 1820, of free blacks, *by the American* colonization Society. It has received some accessions, and the *prospect of its permanency* and growth are flattering The *settlement is called* Liberia.

3. COAST OF GUINEA.

Q. How is the coast of Guinea *bounded?*

A. N. by Sierra Leone and c... a...; E. by c... a...; S. by the c... of c..., or l... g... and a... o...; and W. by the a...

Q. How is the coast of Guinea *subdivided?*

A. Into the Grain coast, Ivory coast, Gold coast, Slave coast, and the kingdoms of Benin and Biafra; together with the kingdoms Ashantee and Dahomey.

1. The *Grain coast* is the most northerly part of Guinea. It is *not distinguished* for any valuable articles of trade, and is little frequented.

2. The *Ivory coast,* which *lies* south, *abounds with* ivory, but is *seldom visited,* owing to the want of harbours. The *inhabitants* are said to be peculiarly savage.

3. The *Gold coast lies* east of the Ivory coast. A greater *trade* has been carried on from this coast than from any other part of Africa. The *principal articles of commerce* are gold and ivory. The *trade* is principally in the *hands of* the English and Dutch. Cape Coast castle, *large town,* is the *chief town*

and *capital* of the English settlements. Elmina, *6th class*, is the *capital* of the *Dutch settlements*.

4. The *Slave coast lies* between the Gold coast and Benin The *country was formerly* in a high state of cultivation, the inhabitants industrious and prosperous. It is *new subject to* the king of Dahomey, whose viceroy exercises a military tyranny over the people. The country no longer flourishes. Great numbers of slaves were formerly exported from this coast.

5. *Benin lies* east of the Gold Coast. But little is known of the country. The *government* is an absolute monarchy. The *inhabitants* are represented as gentle in their manners, and superior to most of the African tribes in agricultural skill.

6. *Biafra is situated* northeast of Benin, and borders upon it, but is almost wholly unknown.

7. *Ashantee is situated* north of the Gold coast. This *kingdom* is the most powerful, civilized, and commercial, of any in Western Africa. The *chief town* and *capital* is Coosmassie, *5th class*. The houses are small, but its palace is a magnificent structure.

8. *Dahomey* is a considerable kingdom, *lying* northeasterly of Ashantee. The *government* is the most despotic known. The *king is regarded* as a superior being, and his subjects consider themselves as his slaves. War is their delight, and human skulls are a favourite ornament in the construction of their palaces and temples.

4. COAST OF CONGO, OR LOWER GUINEA

Q. How is the Coast of Congo *bounded?*

A. N. by the coast of G...; E. by c... A....; S. by s... A...; W. by the A...

Q. How is the territory of Congo, or Lower Guinea *divided?*

A. Into Loango, Congo, Angola, and Benguela.

1. *Loango extends* from about latitude 2° 20′ S. to the river Zaire, a distance of more than 400 miles. This *coast has been visited by* the Portuguese and French almost exclusively, for the purchase of slaves. Loango, *6th class*, is the *chief town* and *capital*.

2. *Congo lies* to the south of Loango, from which it is separated by the river Zaire. From this coast the Portuguese carry on the slave trade. St. Salvador, *5th class*, is the *chief town* and *capital*. It is *situated* in the interior, 150 miles from the sea, on a high mountain. It was formerly called Banza.

3. *Angola lies* immediately south of Congo. The Ports

guese have settlements in this part of Africa. The *chief town* and *capital* is Loando St. Paul, *6th class.*

4 *Benguela lies* south of Angola. The *climate* is very unhealthful. The *inhabitants* are rude and barbarous. Benguela, *small town,* is the *chief town* and *capital.*

Questions.—Map of Africa.

How is Western Africa bounded? What countries does it comprehend? Which are the principal rivers? What range of mountains forms a part of the eastern boundary? How is Senegambia bounded? *Which is the chief town and capital of the Foulahs? Ans. Teemboo. Where is Sierra Leone? Where cape Mesurado? Which is the chief town and capital of Sierra Leone? Ans. Freetown.* How is the coast of Guinea bounded? Into what countries is Guinea divided? *Which are the chief towns and capitals of Ashantee? of Dahomey? of Benin?* What Gulf lies south of Guinea? How is Congo, or Lower Guinea bounded? How is it divided? *Which are the chief towns and capitals of Loango, Congo, Angola, and Benguela?* Which is the principal river on the coast of Congo?

XXX. Southern Africa.

Cape Town.

Q. How is Southern Africa *bounded?*

A. N. by w... A..., c... A..., and E... A...; E. by the I... o...; S. by the s... o...; W. by the A... o...

Q. How is Southern Africa *divided?*

A. Into the Colony of the Cape of Good Hope, and Caffraria.

1 COLONY OF THE CAPE OF GOOD HOPE.

Q. How is the colony of the Cape of Good Hope *bounded?*

A. N. by c...; E. by c... and the I... o...; S. by the s... o...; W. by the A... o...

1. *The face of the country* is diversified by alternate plai and ridges of mountains, which extend across the country fro east to west. The *soil* along the southern coast is deep i fertile. The *interior contains* large tracts of arid desert; t *northern region* is occupied by a plain 300 miles long, and i broad, *called* the Great Karoo. It is nearly destitute of veg tation.

2. The *climate* is hot during the summer months, and parched by a dry wind. In the cold season the settlement deluged with rain; still the *climate* is considered as healthfu and the cape is a celebrated resort for the invalids of India.

3. The *length* of the Colony of the Cape of Good Hope 600 miles; *breadth* on an average, 200; *square miles*, 120,00 *Pop.* 120,000.

Q. Which are the principal *mountains?*

A. They consist of 3 successive ranges of moun-tains, which run parallel with the southern coast.

1. The *first* of those ranges is *called* Langekloff, or Long. pass, and encloses a space between it, and the ocean, varying from 20 to 60 miles in length. *Class* uncertain.

2. The *second range* is *called* Zwarte Berg, or Black moun-tain. *Class* uncertain.

3. The *third range*, which is from 80 to 100 miles north of the second range, is *called* Nieuweldts, or Snowy mountains, *5th class.*

4. *Table mountains, 6th class,* a stupendous mass of naked rock, *lies* behind Cape Town.

Q. Which are the principal *rivers?*

A. Great Fish river, and Orange river.

1. *Fish river, class* uncertain, *rises* in the country of the Hottentots, *flows* southeasterly, and *falls* into the Indian Ocean.

2. *Orange river, 3d class,* the largest river in southern Africa, *rises* in the northeast part of the colony, and *flowing* westerly, *falls* into the Atlantic.

Q. Which are the principal *capes?*

A. Cape Aguillas, Cape of Good Hope, Cape Voltas, and Cape Natal.

Q. Which are the principal *bays?*

A. Table bay, False Bay, Saldanha, and Algoa Bay.

Table Bay *lies* north of the Cape of Good Hope. False Bay *lies* south; Saldanha bay *lies* north of Table bay, and is the most considerable harbour in Southern Africa. Algoa bay *lies* in the southeastern extremity of the colony.

Q. Which is the *chief town* and *capital?*

A. Cape Town.

Cape Town, 6th class, is pleasantly *situated* upwards of 30 miles from the cape, and is an important commercial town. Ships frequently call here in their voyages to, and from India. *Bethesda* and *Bethelsdorp* are missionary stations, the former near Orange river, and the latter near Algoa Bay.—They are chiefly inhabited by Hottentots.

Q. Who are the *Hottentots,* and what is their *character?*

A. They are the aborigines of the country, who are represented as ignorant, stupid, filthy, and sensual; but generally mild, docile, and hospitable.

The *Hottentots consist* of three *classes,* viz. the inhabitants of the colony, who are estimated at 15,000; the *Bosjesmans,* or wild Hottentots, who inhabit the mountainous districts, extending along the northern frontier of the colony. They are excessively ugly, deformed, mischievous, and indolent; and the *Namquas,* who occupy the northwestern coast.

2. CAFFRARIA.

Q. How is Caffraria *bounded?*

A. N. by C... A... and E... A...; E. by the I... O...; S. by the Colony of the Cape; W. by the A... O...

Q. Who are the *inhabitants,* and what is their *character?*

A. Caffres, who are represented as bold, intelligent, active, and ingenious.

The *Caffres consist* of several tribes, *some* of which are similar to the Hottentots. Latakoo, *large town,* is the *chief town* and *capital of the Bushman tribe;* Kurreechane, *6th class,* the *chief town* and *capital* of the *Marootze tribe.*

Questions.—Map of Africa.

How is South Africa bounded? How is it divided? What desert lies between South Africa and Benguela? Ans. Cimbebas.—Which are the principal capes? What range of mountains lies north of Cape Town? What river north of the Snowy mountains?—Which is the chief town and capital of the colony? *Where is Latakoo? What is the latitude and longitude of Cape of Good Hope? Which extends farthest south, Africa or New Holland? Africa or South America?*

XV. Eastern Africa.

Method of Carrying the Nobility.

Q. How is Eastern Africa *bounded?*

A. N. by E..., from which it is separated by the tropic of Cancer; E. by the R... S..., Strait of B..., and I... O...; S. by S... A..., from which it is separated by the tropic of Capricorn; W. by C... A...

1. The *face of the country* of those portions of Eastern Africa, which are known, is considerably diversified, consisting of plains, deserts, marshes, and occasional mountainous regions.

2. The *soil* in some parts is very fertile, but extensive regions of barren waste, occur.

3. The *climate* also varies with the situation, being in some places very healthful, and in others subject to the diseases of tropical climates; the country is also subject to the simoon, or poisonous wind from the desert.

Q. Which are the principal *mountains?*

A. The mountains of Lupata, which run parallel with the coast of Zanguebar; and the mountains of Abyssinia.

1. The *height* of the mountains of Lupata is unknown, and even the existence of such a range is doubted by some.

2. Of the *mountains of Abyssinia*, the *principal* are Geesh, a volcano, *4th class*, the mountains of Amid-Amid, *4th class*, Lamalmon and Gondar, *5th class*.

Q. Which are the principal *lakes?*

A. Maravi and Dembea.

Maravi lies about 300 miles from the coast of Mozambique, west of the mountains of Lupata. It is 300 miles *long*, and 30 *broad*. *Dembea* is a large lake in the western part of Abyssinia, supposed to be 450 miles in *circumference*.

Q. Which are the principal *rivers?*

A. The Bahr-el-Azrek, or Blue river, and the Tcazze.

Both these rivers rise in Abyssinia, and passing through Sennaar, at length fall into the Nile.

Q. How is Eastern Africa *divided?*

A. Into Nubia, Sennaar, Abyssinia, and the countries south of Abyssinia.

1. NUBIA.

Q. How is Nubia *bounded?*

A. N. by E...; E. by the R... S...; S. by S...; W. by C... A...

1 The *face of the country*, excepting on the banks of the Nile, is sandy and rocky. The *soil* on the banks of the Nile is rich and fertile.

2. The *climate* is dry and hot, but healthful.

Q. Which is the principal *river?*

A. The Nile.

Q. Which is the *chief town* and *capital?*

A. Dongola.

1. Dongola, *small town*, is *situated* on the Nile, 280 miles south of Syene. It is meanly built, and in a state of decay.

2. The *other principal town* is Suakem, on the Red Sea.

Q. Which are the principal *productions?*

A. Dhurra, a kind of grain, barley, beans, and tobacco.

Q. What is the *character* of the inhabitants?

A. They are chiefly Arabs, possessing the common traits of Arabs; although some of them are more stationary and pay greater attention to agriculture.

Q. What is the *government?*

A. An absolute monarchy.

The various tribes are governed by chiefs, who are described as very violent and arbitrary.

Q. What *curiosities* does Nubia contain?

A. Many magnificent ruins of ancient cities.

18

The *most remarkable* of these is the temple of Ebsamba which was cut out of a solid rock, and now remains entire, its front is 117 feet long, and 86 high.

Q. Which is the principal *article of trade?*

A. Slaves, who are brought from the interior of Africa, and conveyed into Egypt and Arabia.

The *number* of slaves thus sold is estimated at 5000 annually, of whom 2500 are for Arabia, and 1500 for Egypt.

2. SENNAAR.

Q. How is Sennaar *bounded?*

A. N. by N...; E. and S. by A...; W. by D...

The *face of the country* is generally level. The *soil* between the Nile and the Tcazze is very fertile; other extensive tracts are perpetually desert. *Pop.* 2,000,000.

Q. Which is the *chief town* and *capital?*

A. Sennaar.

Sennaar, 3d class, is *situated* on the Bahr-el-Azrek, about 800 miles above its junction with the Abiad, or main branch of the Nile.

Q. Which are the principal *productions?*

A. Dhurra, rice, grain, melons, tobacco, and sugar cane.

Q. What is the *government?*

A. A despotic monarchy.

Q. Which are the principal *articles of trade?*

A. Gold dust, ivory, and slaves.

3. ABYSSINIA.

Q. How is Abyssinia *bounded?*

A. N. by S...; E. by the R... S...; S. by A... and unknown regions; W. by S... and unknown regions.

The *face of the country* is mountainous. The *soil* is very fertile. The *climate* in the elevated regions, is cool and salubrious; in the valleys, or in the vicinity of marshes, it is hot, and the air suffocating. *Length*, estimated at 800 miles; *breadth*, 600; *square miles*, 450,000. *Pop.* from 1 to 4,000,000

Q. Which are the principal *mountains?*

A. The mountains of Geesh, Amid-Amid, Lamalmon, and Gondar.

Geesh is a volcano of the 4th class, near the supposed

ource of the Nile. *Amid-Amid, 4th class,* is *supposed to be a* branch of the mountains of the Moon. *Lamalmon, probably 5th class, bars* the entrance of the Red Sea. *Gondar, 5th class,* is not far from the city of Gondar.

Q. Which are the principal *rivers?*

A. The Bahr-el-Azrek, and the Tcazze.

The *Bahr-el-Azrek,* or Blue river, *rises* from two fountains near Geesh, passing through the lake of Dembea, it *runs* in a semi-circular course, *gradually winding* to the north, which direction it pursues till it reaches latitude 10, when it *unites* with the principal branch of the Nile. The Tcazze *rises* about 150 miles east of Gondar, and *running* northwesterly, through Sennaar, *falls* into the Nile in Nubia, in lat. 17° 45'.

Q. Which is the principal *lake?*

A. Dembea.

Q. Which is the *chief town* and *capital?*

A. Gondar.

1. *Gondar, 4th class,* is *situated* on a hill of considerable height, near lake Dembea. Its appearance is mean, but it has some fine buildings.

2. The *other principal towns,* are Axum, *small town,* the ancient capital, 150 miles N. E. of Gondar; Masuah, *small town,* the only place of foreign trade; and Adowa, *large town,* the seat of trade for the interior.

Q. What is the *religion* of the Abyssinians?

A. They profess Christianity, but are extremely ignorant.

Q. What is the *character* of the Abyssinians?

A. They are in a low state of civilization, and in many of their customs are barbarous and brutal.

Q. Which are the principal *productions?*

A. Wheat, balsam, and myrrh.

Q. Which are the principal *articles of manufacture?*

A. Cotton cloths, arms, and instruments of iron.

Q. Which are the principal articles of *export* and *import?*

A. The *exports* are gold, ivory, and slaves; the *imports* are lead, block-tin, Persian carpets, raw silk, and broadcloths.

Q. What is the *government?*

A. A despotic monarchy.

The Gallas, a savage people, on the south, have recently overrun Abyssinia, and now possess Gondar.

4. COUNTRIES SOUTH OF ABYSSINIA.

Q Which are the *principal countries* in Eastern Africa, south of Abyssinia ?

A. Adel, Berbea, the coasts of Ajan, of Zanguebar, of Mozambique, of Sofala, and Mocaranga.

1. *Adel lies* on the coast southeast of Abyssinia. The *inhabitants* are Mahometans, and the *government* despotic The *chief town and capital* is Zeila.

2. *Berbera extends* from Adel to Cape Guardafui. It *abounds* in gums, myrrh, and frankincense, more than any other part of the world. Berbera is the *chief town* and *capital*. The *inhabitants* appear to be considerably civilized and commercial.

3. The *coast of Ajan extends* from Cape Guardafui to the river Magadoxa. The *country* is sandy, flat, and barren. The *inhabitants* are Mahometans, who carry on a considerable trade in ambergris, and gold.

4. The *coast of Zanguebar extends* from the river Magadoxa to cape Delgado. It is *divided* into several kingdoms. The *country* is marshy and unhealthful. The *inhabitants* are partly pagans and partly Mahometans. The *governments* are absolute monarchies.

5. The *coast of Mozambique extends* from cape Delgado to the river Cuama. The *soil* is fertile. The *country* is *subject* to the Portuguese. It *abounds* in gold, and in fierce and destructive wild beasts. Mozambique, *small town*, is the *chief town* and *capital*.

6. The *coast of Sofala extends* from the river Cuama to Cape Corrientes. The *rivers* which water this country, are the Cuama, Sofala, and Inhambane ; all of which *flow* easterly into the Indian Ocean. The *principal settlements* are Sena, *small town*, on the river Cuama, 250 miles from its mouth ; and Sofala, *small town*, *situated* near the mouth of the river of the same name.

7. *Mocaranga* is a kingdom *situated* in the interior, and appears to be one of the most powerful in this country. Zimbao is the *capital*.

Questions.—Map of Africa.

How is Eastern Africa bounded ? Which are the principal countries ? How is Nubia situated ? What river passes through Nubia ? Which is the chief town and capital ? What country lies south of Nubia ? Which is the chief town and capital of Sennaar ? What country lies south of Sennaar? What great river rises in Abyssinia? What is the chief town and capital? Mention the countries in Eastern Africa south of Abyssinia. Where are the mountains Lupata? Where is the island of Maravi?

V. Central Africa.

Hottentots.

Q. How is Central Africa *bounded?*

A. N. by n... a...; E. by e... a...; S. by s... a...; W. by w... a...

Q. Which are the principal *rivers* of Central Africa?

A. The Niger and the Wad-el-Gazel.

The *Niger rises* in the western part of the Mountains of the Moon, and *flows* easterly; its *termination* is not known The *Wad-el-Gazel rises* west of the sources of the Nile, and *flows* northerly, till it is *lost* in the sands of the desert.

Q. Which are the principal *mountains?*

A. The Mountains of the Moon.

These mountains are supposed to *extend from* Nigritia through Abyssinia *to* the Indian sea.

Q. What is the *character* of the inhabitants?

A. They consist of Negroes, Moors, and Arabs, and possess the traits of character peculiar to these respective tribes.

Q. Which are the principal *articles of export and import?*

A. The exports are, gold, slaves, ostrich feathers,

and tiger skins; the articles of *import* are **East** India goods, fire arms, sabres, knives, looking-glasses, and red worsted caps.

Q. What is the *government* of the respective tribes?

A. A despotic monarchy.

Q. How is Central Africa *divided?*

A. Into Nigritia, and Lower Ethiopia.

The country called *Lower Ethiopia* has been so little explored that no account of it can be given in geography.

NIGRITIA.

Q. How is Nigritia *bounded?*

A. N. by N... A..., from which it is separated by the Saharadese; E. by E... A...; S. by L... E... and W... A...; W. by W... A...

Q. Which are the principal *kingdoms* of Nigritia?

A. Dar'fur, Bornou', Fezzan, Kassina, or Kashna, Houssa, (Hoo'sah) Tombuctoo, and Bambarra.

1. *Darfur is situated* to the westward of Abyssinia; considerable attention is paid by the inhabitants to agriculture. *Pop.* is estimated at 200,000. The *religion* is Mahomedan; the *government* despotic. The *chief town* and *capital is* Cobbe, *large town.*

2. *Bornou is* S. W. of L. Tchad. It is the most powerful kingdom of Central Africa. The *climate* is hot; the *soil* fertile; the *productions* are grain, maize, rice, together with the fruits of tropical climates. The *religion* is partly Mahomedan; the *government* an absolute monarchy. The *capital* is Bornou, *2d class.*

3. *Fezzan lies* south of Tripoli, and forms, as it were, a great island in the midst of the Sahara. It is 300 miles *long,* and 200 *broad.* The *heat of summer* is intense; the *winters* are accompanied by cold and bleak winds. *Mourzouk, 6th class,* is the *chief town and capital,* and is the grand depot for the immense commerce carried on between Northern and Central Africa.

4. *Kashna lies* west of L. Tchad. The *people* are said to have some skill in the arts. The *face of the country* is generally level, consisting of land of great fertility, interspersed with arid wastes. It is said to contain 1000 *towns* and *villages.* Kashna, *class* uncertain, is the *chief town* and *capital.*

5. *Houssa lies* west of Kashna. The *people* are the most

atelligent of any in the interior of Africa. *Agriculture* is in high state of perfection. Houssa, the *chief town*, is of the 3d *class ;* Sackatoo is the capital.

6. *Tombuctoo lies* west of Houssa on both sides of the Niger. It is one of the most powerful kingdoms in Central Africa. Although the accounts respecting it are extremely vague, yet the country is thought to be considerably civilized and opulent. The *chief town* and *capital* is Tombuctoo, probably of the 3*d class.* It is *situated* not far from the Niger, about 80 days' journey from Tripoli.

6. *Bambarra lies* south of Tombuctoo. It is traversed from west to east by the Niger, and is generally very fertile. The *government* is a despotic monarchy. Sego, 5*th class,* is the *chief town and capital.* It is *situated* on both sides of the Niger, and is surrounded by high mud walls.

Questions.—*Map of the World.*

How is Central Africa bounded? Which are the two great divisions of Central Africa? What chain of mountains separate them? Which are the principal rivers of C. A.? Which are the principal countries of Nigritia? *Which is the chief town and capital of Darfur? Of Bornou? Of Kassina? Of Fezzan? Of Houssa? Of Tombuctoo? Of Bambarra? What is the latitude and longitude of Tombuctoo? How far in a right line is Tombuctoo east of the mouth of the Senegal? Ans. 1200 miles. How far southwest of Cairo? Ans. About 2000.*

VI. African Islands.

Peak of Teneriffe.

Q. Which are the principal *Islands* on the *western coast of* Africa?

A. The Azo'res, Madeira, and Cape Verde Islands, belonging to Portugal ; the Canaries belonging to Spain, and St. Helena to Great Britain.

Q. What is the *number* and *situation* of the Azores ?

A. They consist of nine islands, situated about half way between Europe and America.

1. The *principal of the Azores* are St. Michael, Fayal, and Tercera.

2. The *soil* of these islands is fertile, and the *climate* healthful. The principal *productions* are oranges, lemons, corn and wine. The islands are *subject* to earthquakes. The *capital* of Tercera is Angra ; Pontadel Gada, *large town*, is the *capital* of St. Michael.

Q. What is the *situation* of Madeira ?

A. It lies on the north-western coast of Africa.

This island is 54 miles long, and 21 broad. It is very fertile, and is *celebrated* for its wine, which is exported to all parts of the world. The *population* is 80,000. Funchal, 6th *class*, is the *chief town and capital*.

Q. What is the *number* and *situation* of the Cape Verde Islands ?

A. They are 10 in number, and lie 390 miles W. of Cape Verde.

These islands are mountainous ; but the *climate* is very hot and unhealthy. *Porto Praya* is the chief town.

Q. What is the *number* and *situation* of the Canaries ?

A. They are 14 in number, lying south of Madeira, near the African coast.

The principal are Teneriffe, Grand Canary and Palma. The *climate* is delightful. The *productions* are wine, sugar, grain and fruits. The *peak of Teneriffe* is celebrated.

Q. What is the *situation* of St. Helena ?

A. It lies off the south-western coast of Africa.

This island is 10 miles *long*, and 6 *broad*, and is *surrounded* by rocky precipices or bluffs, some of which are 1600 feet high. The *island is famous for* having been the prison of Napoleon Bonaparte, from August 1815, to the time of his death, March 5, 1821.

Q. Which are the *principal islands* on the *eastern* coast of Africa ?

A. Madagascar, and the Comoro Islands, belonging to the natives ; Bourbon, belonging to France, and Mauritius, belonging to Great Britain.

Q. What is the *situation* of Madagascar ?

A. It lies on the south-eastern coast of Africa.

Madagascar is about 900 miles *long*, and is one of the large islands on the globe ; the *climate* is very temperate and thelous. It *produces* rice, potatoes, sugar cane, &c. The inhabitants are estimated from 1 to 4,000,000; but they are very little civilized.

Q. What is the *situation* of the COMORA ISLANDS?

A. They lie north of Madagascar.

These islands are *inhabited only by* natives; they are very fertile, and are well stocked with cattle, birds, and wild animals.

Q. What is the *situation* of the Isle of BOURBON?

A. It lies east of Madagascar.

It enjoys a fine *climate*, and *produces* corn, coffee, and rice. Its *population* is 100,000.

Q. What is the *situation* of MAURITIUS, or ISLE OF FRANCE?

A. It lies further east than Bourbon.

It is strongly fortified, and was formerly the chief naval station of the French in the Indian seas; but was taken by the British in 1810.

Questions.—Map of the World.

Where are the Azores? What is the situation of Madeira? What islands lie west of Cape de Verde? Which way from Madeira do the Canaries lie? What is the situation of St. Helena? Where is the large island of Madagascar? What channel separates it from the continent? Which way from Madagascar do you find the Comoro Islands? Which way the Isle of Bourbon? Which way from the Isle of Bourbon is Mauritius, or isle of France?

OCEANICA.

Branch of the Bread Fruit Tree.

Q. Which is the 5th *division* of the World?

A. Oceanica; a term meaning the islands of the ocean. It includes the numerous clusters of islands,

19

lying east, south, and southeast of Asia, togethe
with other clusters in the Pacific ocean, usually calle
Polynesia.

Q. How is Oceanica *bounded?*

A. N. by the 40th degree of north latitude ; S. by
the 50th of south lat.—It extends from the 92d de
gree of east longitude, to the 158th degree of west
longitude.

The following are the boundaries, according to Malte Brun :—

By a line drawn from the northern point of Sumatra through the
strait of Malacca, China sea, between Formosa, and the Philippine
isles, to the point where the 40th degree of N. lat. and the 152d degree of E. lon. intersect each other ; thence along the parallel of
lat. to the 158th degree of W. lon. ; thence southerly through the
108th degree of W. longitude, on the equator, to the 50th degree of
S. lat., from which it proceeds on this parallel to the 92d degree of
E. lon. and thence northerly to the point of starting.

1. Oceanica, as thus marked, *embraces* 90 *degrees* of lat.
40 degrees *north*, and 50 degrees *south* lat. ; or about 6250
miles : and 160 *degrees of lon.*—88 degrees of *east*, and 72
degrees of *west* lon. ; or about 11,000 *miles.*

2. The *climate* of many of the islands of Oceanica, is
delightful ; that of others, especially of such as are low and
marshy, is unhealthful.

3. The *soil* is in general, very fertile, yielding the richest
fruits, spices, and gums.

Q. What can you say of the *mountains* of these islands ?

A. They have generally a marked direction from
north to south, at the same time bending about the
middle of the ranges, from west to east.

On some of these islands lofty mountains are found, which
would seem to indicate that the islands are the fragments of
a great continent ; others are volcanic, and not a few are
very low, having their origin, it is supposed, in coral reefs.

Q. What great *curiosity* exists in the seas of Oceanica ?

A. Coral reefs.

1. These reefs are *supposed to be the work* of very small
animals, called *zoophytes.*

2. They often extend from island to island, rendering the
navigation of these seas exceedingly dangerous. In some
cases they rise from the bottom of the ocean, 6 or 700 feet,
and form a foundation, upon which, sand being washed, and
seeds being brougnt by the birds and winds, in process of
time, islands have been formed.

Q. Which are some of the principal *straits*?

A. The straits of Sunda, Malacca, Macassar, Endeavour, Bass's, Dampier's, and Cook's.

1. The *strait of Sunda separates* Sumatra from Java.
2. The *strait of Malacca*, the Malay coast from Sumatra.
3. The *strait of Macassar*, Borneo from Celebes.
4. *Endeavour strait* separates New Guinea from New Holland.
5. *Bass's strait separates* New Holland from Van Diemen's
land.
6. *Dampier's strait separates* New Guinea from New
Britain.
7. *Cook's strait separates* the two islands of New Zealand.

Q. What is the *government* of these islands?

A. It is generally a hereditary, despotic monarchy

Q. How is Oceanica *divided?*

A. Into North Western—Central—and Eastern
Oceanica, or Polynesia.

Questions.—*Map of the World.*

How is Oceanica bounded? (*See Geog.* p. 218.) *Through how
many degrees of lat. does it extend? How many degrees of longitude?* Which is the largest island in Oceanica? Which is the 2d?
Which the 3d? What strait separates Sumatra from Java? What
separates the Malay coast from Sumatra? What Borneo from Celebes? What New Guinea from N. Holland? What N. Holland from
Van Diemen's Land? What New Guinea from New Britain?
What separates the two islands of New Zealand? How is Oceanica
divided, and which are the principal islands belonging to each
division?

I. North Western Oceanica.

Q. What *islands* are included in NORTH WESTERN OCEANICA?

A. All the islands between Malacca and the China
sea on the north, and New Guinea and New Holland
on the south—the principal of which are Sumatra,
Java, and Borneo, called Sunda isles—the Philippine islands, and the Moluccas, or spice islands.

1. ISLES OF SUNDA.

Q. Which are the *Sunda Isles*, and how are they *situated?*

A. They are Sumatra, Java, and Borneo, lying
south of Malacca, and the China sea.

1. *Sumatra*. The *length* of Sumatra is about 1000 miles from northwest to southeast; the *breadth*, varies from 55 to 235 miles. *Sq. miles* about 160,000. *Pop.* 4,500,000. A *chain of mountains runs* through the island near the western coast; Mt. Ophic, *4th class*, is the *highest peak*, and is directly under the equator. The *climate* is cool, notwithstanding the country lies under the equator; the *western coast* is low and sickly. The *principal productions* are rice, pepper, cotton, camphor, and benzoin; rattans grow here: gold abounds. The *tame animals* are horses, cows, sheep; the *wild animals* are the elephant, tiger, rhinoceros, black bear, otter, and wild boar. Of birds, the most *beautiful*, and perhaps the richest of all the feathered race, is the coo-ow, or argoa pheasant. The country is *divided* into several kingdoms. The *interior is inhabited* by a race of cannibals; the *coast* is possessed by Malays. The *governments* are generally hereditary despotisms; the *religion* is pagan. The only English *settlement on the island*, is Bencoolen.

2. *Java*. The *length* of Java is 690 miles; *breadth* varying from 80 to 140; *square miles* 51,000; *Pop.* about 5,000,000. The *climate* is in many parts extremely unhealthful; the *soil* is very rich. A *range of mountains* runs through the island from E. to W. Many of the summits are volcanic. The *principal productions* are rice, sugar, pepper, and coffee; they *export* indigo, teak timber, spices brought from the Moluccas, tin from Banca. The *inhabitants* are Javanese and Malays, who are of the Mahomedan *religion*. The Javanese are indolent, superstitious, revengeful, and slow of understanding; but are remarkable for their candour and veracity. The *government* is a hereditary despotism. The *island belongs* to the Dutch. Batavia, *5th class*, is the *capital*. The city was *founded* in 1619; *taken by the British* in 1811, but *restored* to the Dutch, in 1816.

3. *Borneo*. The *length* of Borneo is about 800 miles; *breadth* 700; *square miles* about 390,000. *Pop.* uncertain, yet commonly reckoned at 3,000,000. The *face of the country* on the coasts for some distance inland, is low and marshy; the *interior* is partly mountainous; the *climate* is unhealthful. *Earthquakes* are frequent. It has many considerable *rivers*. The *productions* are rice, pepper, camphor, and all the fruit trees of India. *Diamonds* of great value are found here. The *bird* of Paradise is common in the island; also monkies of the largest kind, and the Ourang Outang, which so strongly resembles the human species. The *inhabitants*

on the coast are Malays, Javanese, Bugis, or natives of Celebes, all of the Mahometan *religion*. The *interior of the
country is divided* into independent tribes, the *government is*
despotic. Borneo, on the N. W. part of the island, is the
chief town and *capital*.

2. PHILIPPINE ISLANDS.

Q. Which are the principal of the *Philippine islands*, and
how are they *situated?*

A. Lucon, (Lu-son') and Mindanao are the principal; which, with the others of the cluster, lie N. E.
of Borneo.

1. The *number* of the Philippine islands is said to be 1200,
of which 500 or 600 are of importance. These islands
abound in marshes, mossy grounds, and lakes. *Earthquakes,*
frequently occur. *Mountains,* rise to some height, and are
full of volcanoes. The islands are *subject to* great inunda-
tions from rain. *Pop.* 3,800,000. *The country belongs to*
Spain, but the principal islands are in the hands of independent tribes,

2. *Lucon.* The *length* of Lucon, (Luson,) the largest, is
about 400 miles from N. to S. and from 90 to 120 in *breadth;*
square miles about 65,000; *Pop.* estimated at 1,000,000. The
chief town and *capital* is Manilla, 5th class.

3. *Mindanao.* The *length* of Mindanao, which lies southeast 600 miles from Lucon is 300 miles; *breadth,* 150. The
chief town and *capital* is of the same name.

3. MOLUCCAS, OR SPICE ISLANDS.

Q. How are the *Moluccas,* or *Spice islands, situated,* and
which are the *principal?*

A. They lie east of Borneo and Java, and south
of the Philippines, extending to the immediate neighbourhood of New Guinea and New Holland. The
principal of these islands are Celebes, Gilolo, Ceram,
Amboyna, and Banda.

1. *Celebes.* The *length* of Celebes is 500 miles; *breadth*
200; *square miles* 90,000. *Pop.* 3,000,000. The island contains several *volcanoes* in a state of activity. The *productions*
of the island are rice, cotton, pepper, cloves, nutmegs, and
the like. The *religion* is Mahometan; the *governments* of
the island are aristocracies combined with elective monarchy.
The island *belongs to* the Dutch, who have a flourishing town
on the S. W. coast, called Macassar, 3d class.

2. *Gilolo.* *Gilolo* is 230 miles *long,* or about 600 in *circum-*

19*

ference. It produces sago, formed from the pith of a large tree. One tree when 15 years old, will sometimes yield 5 or 600 lbs.; the average is supposed to be 300 lbs. The island also yields bread fruit. This is so large as to contain about two quarts of meal within the shell. The meal makes good bread; it supports vast numbers in the islands of the Indian and Pacific ocean.

3. Ceram. Ceram is 186 miles long, and 36 broad. It is the most distinguished of all the neighbouring islands for yielding sago.

4. Amboyna. Amboyna is 55 miles long. It is particularly celebrated for its cloves, of which 650,000 lbs. are produced annually. The Clove tree grows to the height of 40 or 50 feet. The cloves are gathered twice a year. The inhabitants are indolent, effeminate, and licentious. They are partly Mahometan, Roman Catholic, and Protestant. Missionaries are supported here by the London and Baptist Missionary Societies. The number of inhabitants in the island is between 40 and 50,000. Amboyna, large town, a Dutch settlement, is the chief town and capital.

5. Banda Isles. The Banda isles are ten in number, and are celebrated for producing great quantities of nutmegs. The islands are of volcanic origin, and are considered very unhealthful.

6. Timorian Isles. To the above islands may be added another group, sometimes called "the great Timorian chain," from Timor, the largest of them, around which they lie. In this group is Sumbawa, celebrated for its volcano, Tamboro, which in 1815 threw out ashes in such quantities as to cause total darkness on the island for 22 hours.

Questions.—Map of Oceanica.

What islands does North Western Oceanica include? (p. 219.) Which are the principal of the Sunda Isles? How does Sumatra lie in respect to Malacca? What strait separates them? What is the chief town and capital of Sumatra? Which way from Sumatra is Java? What strait separates them? What is the chief town and capital of Java? Which way from Java is Borneo? What is the chief town and capital of Borneo? Which way from Borneo are the Philippine Islands? Which are the two principal islands? What is the chief town and capital of Lucon? How are the Spice Islands situated? Which is the largest? Which way from Celebes is Gilolo? What is the chief town and capital of Celebes? Where is Sumbawa? Where Timor? What islands in North Western Oceanica does the equator cross?

II. Central Oceanica.

Q. What islands are included in CENTRAL OCEANICA?

A. New Holland and the numerous islands round

it, viz. Van Diemen's Land, New Guinea, New Britain, New Ireland, Solomon's Island, Louisiade, New Caledonia, New Hebrides, New Zealand, and some others.

Central Oceanica is estimated to contain 3,500,000 *square miles*, and a *population* of 5,000,000.

1. NEW HOLLAND, AND VAN DIEMEN'S LAND.

Q. What is the *situation* of *New-Holland?*

A. It lies southeast of North Western Oceanica.

1. The *length* of New Holland from north to south is about 2600 miles; *breadth* from east to west, 2000. From its extent it has *sometimes been called* a continent.

2. The *face of the country* on the coasts, is extremely diversified. A *chain of mountains runs* in a direction parallel with the eastern coast, at a distance of from 500 to 800 miles. —The *highest peaks* belong to the 6*th class*.

3. None of the *rivers* on this coast have the appearance of a long course. Hawkesbury river, which is navigable for 140 miles, is the *principal*. It *falls* into the sea at Broken bay, north of Port Jackson.

4. The *seasons* in New Holland correspond with those of the south of Africa and America; its summer answering to our winter, and its spring to our autumn.

5. The *native productions* which furnish food for man are very few; but the various kinds of grain, fruit, and vegetables, which grow in England, have been introduced, and yield abundantly.

6. The *original inhabitants* of the country appear to be in the lowest rank of intellectual being. They are of low stature, and ill shaped; their complexion is nearly black, and their appearance extremely filthy and disgusting. It is said they have no *religion*, and scarcely any idea of a future state.

7. An English *colony was established* on the eastern part of Botany Bay, in 1788; but soon after, the settlement was removed to Port Jackson, 12 miles north. To this colony great numbers of criminals have been banished from Great Britain. The colony is very flourishing, and many of the banished have become virtuous, and respectable citizens. Sidney, *large town*, is the *chief town* and *capital*. The *other principal settlements* are Paramatta, 15 miles from Sidney; and Windsor, on the Hawkesbury, 35 miles from Sidney.

Q What is the *situation* of *Van Diemen's Land?*

A. It is separated from New Holland by Bass strait, which is 100 miles wide.

This island is about 170 miles *long*, and 150 *broad*. The *climate* is very healthful, and the *soil* fertile. The isle contains several *mountains* of considerable elevation. The *Aborigines* of this island, bear a great resemblance to those of New Holland. The British have planted a *colony* on the southeast part of this island, called *Hobert Town*. The colony is flourishing.

2. NEW GUINEA.

Q. What is the *situation* of *New Guinea?*

A. It lies northeast of New Holland, from which it is separated by Endeavour strait, 15 miles broad.

1. This country is often *called* Papua. Its *length* is about 1200 miles, and its *breadth* from 15 to 360.

2. But little is *known* either of the country or its inhabitants. The *coasts* are generally high, and some of the mountains of the interior, are said to be 2000 feet high. The *inhabitants* are negroes. Their *appearance* is hideous, and their disposition barbarous.

3. NEW BRITAIN AND NEW IRELAND.

Q. What is the *situation* of *New Britain* and *New Ireland!*

A. They lie east of New Guinea, from which the former is separated by Dampier's strait.

The *face of the country*, particularly of New Ireland, is mountainous. The *soil* is fertile. The *inhabitants*, which generally resemble those of New Guinea, are very numerous; those of New Ireland are said to be very warlike. Some of their *canoes* are 80 feet long, and made of a single tree.

4. LOUISIADE.

Q. What is the *situation* of the *Archipelago of Louisiade?*

A. It lies northeast of New Guinea.

This Archipelago *extends* 400 miles in *length*, and 160, where widest, in *breadth*. It *consists of* a number of islands of different sizes, though formerly esteemed a single island. Some of the islands are fertile and populous, and the *inhabitants* warlike, and barbarous.

5. SOLOMON'S ISLES.

Q. What is the *situation* of *Solomon's Isles?*

A. They are situated east of the Archipelago of Louisiade.

The number of *principal islands* is supposed to be 18, from 50 to 300 leagues in circumference. The *air* of these islands is salubrious, *soil* fertile, and the *inhabitants* numerous.

6. NEW CALEDONIA AND NEW HEBRIDES.

Q. What is the *situation* of *New Caledonia* and *New Hebrides?*

A. They are situated southeast of Solomon's Isles.

1. New Caledonia is a large island, being about 220 miles *long*, and 50 *broad*. It is, however, barren, and little known.

2. The *New Hebrides consists of* numerous clusters of islands. They are in general mountainous, and abound in wood and water. The principal *productions* are bread fruit, cocoanuts, plantains, yams, and sugar canes. The *inhabitants* appear civil and hospitable.

7. NEW ZEALAND.

Q. What is the *situation* of *New Zealand?*

A. It lies southeast of New Holland.

1. *New Zealand consists of* two islands *separated* by Cook's strait, which is 12 or 15 miles wide. The *northern* island is 436 miles *long*, and from 60 to 180 *broad*. The *southern* island is 360 miles *long*, and at an average, 100 *broad*.

2. Almost the whole of the northern island is well fitted for cultivation. The southern island is represented by Capt. Cook, as apparently barren. The *climate* is mild and salubrious. The *inhabitants* are said to be tall, sagacious, and intelligent ; but warlike and ferocious. The *government* of the Aborigines is despotic. Several missionary stations have been established, and the success of the missionaries, notwithstanding many discouragements, has been remarkable.

Questions.—Map of Oceanica.

What islands does Central Oceanica include? (p. 222.) Which island is the largest? Which next? What gulf on the N. of New Holland? What names are given to the northern, eastern, and western coasts of N. Holland? What strait separates N. Holland from N. Guinea? What from Van Diemen's Land? Which are the principal capes? What is the chief town and capital? How situated? Which way from New Holland is Van Diemen's Land? Which is the chief town and capital? Where is New Guinea? Which way from N. Guinea is N. Britain? What strait separates them? Where is N. Ireland? What is the situation of Louisiade? In what direction from Louisiade are the Solomon's Isles? Where are New Caledonia and New Hebrides? What is the situation of New Zealand? What strait separates the two islands?

XX. Eastern Oceanica, or Polynesia.

Q. What *islands* are included in *Eastern Oceanica, or Polynesia?*

A. The various groups of small islands, which cover the Pacific from the Pelew islands and the Ladrones to Easter Island, and the Sandwich Isles, the principal of which besides the above, are the Carolines, the Friendly, the Navigators', the Society, and the Marquesas Islands.

1. PELEW ISLANDS.

Q. What is the *number* and *situation* of the *Pelew Islands?*

A. They are 18 in number, and lie E. of the Philippines.

These islands are *often visited* by ships, as places of refreshment. The *inhabitants* are represented as a well formed, amiable, gay, and hospitable people; but without any religion. Fish form their *chief subsistence*, although bread fruit, and cocoas seem to be abundant.

2. LADRONES.

Q. What is the *number* and *situation* of the *Ladrones?*

A. They are 16 in number, and lie N. E. of the Pelew Islands.

These islands *are also called,* Isles of Robbers, being inhabited by a people given to piracy. They *occupy a space* of 450 miles in extent. The *climate* is generally serene, and salubrious, although subject to violent hurricanes. The *inhabitants* are tall, robust, active, and ingenious; but ignorant and superstitious; they have *canoes*, which, with a side wind, will sail at the rate of 20 miles an hour.

3. CAROLINES.

Q. What is the *number* and *situation* of the *Carolines?*

A. They are supposed to be from 30 to 80 in number, and lie E. of the Pelew islands.

But little is known about these islands. Their *climate*, however, is represented as agreeable, and the *soil* fertile. The *inhabitants*, who are very numerous, resemble those of the Philippine islands. They have neither temples nor idols, nor the least *appearance of religious worship*. The *government* is monarchical.

4. FRIENDLY ISLES.

Q. What is the *number* and *situation* of the *Friendly Isles?*

A. They are about 150 in number and lie E. of the New Hebrides.

These islands are in general fertile, and *abounds* in cocoa-nut and bread fruit trees, plantains, sugar canes, and yams. The *population* is supposed to be about 200,000. From their apparent hospitality and kindness, Capt. Cook gave to these islands the name they bear; but subsequent visiters represent them as cruel and ferocious. Of these islands, Ton'gataboo is the *largest*, being 60 miles in circumference; the inhabitants of this island *sacrifice* many human victims, and practise cannibalism.

5. NAVIGATORS' ISLANDS.

Q. What is the *number* and *situation* of the *Navigators' Islands?*

A. They are 10 in number, and lie N. E. of the Friendly Islands.

These islands *received their name from* the admirable degree of skill, displayed by the inhabitants in the management of their canoes. The islands *consist of* high lands, with a fertile *soil;* they *abound in* a variety of provisions, consisting of swine, pigeons, and fruit. The *groves produce* bread fruit, cocoanuts, bananas, guavas, and oranges. The *inhabitants* are represented as an ingenious and industrious people, but exceedingly ferocious.

6. SOCIETY ISLANDS.

Q. What is the *number* and *situation* of the *Society Islands?*

A. They are 60 or 70 in number, and lie east of the Friendly Isles.

1. Of these islands, Otaheite is the *largest*, being in its whole circumference about 120 miles. All the *vegetable species* peculiar to Oceanica, *grow* in Otaheite in abundance, and of the best quality. The *inhabitants* are of a pale mahogany colour, with fine black hair, and eyes, they are mild, affable, and polite. In 1815, the inhabitants of several of these islands *renounced idolatry*, and embraced christianity.

2. Southeast of this cluster is *Pitcairn's island*, a small island, settled by the mutineers of the English ship Bounty. Their *descendants* are an amiable and interesting people.

7. MARQUESAS ISLANDS.

Q. What is the *number* and *situation* of the *Marquesas Islands?*

A. They are five in number, and lie N. E. of the Society Islands.

The *trees, plants*, and *other productions* of these islands strongly resemble those of the Society Isles. The *inhabitants*, vaguely *estimated at* 50,000, are said to *surpass* all others in this sea in the symmetry of their shape, and the regularity of their features.

8. EASTER ISLAND.

Q. What is the *situation* of *Easter Island?*

A. It lies east of Pitcairn's island, and is the most eastern in Oceanica.

This island is small, containing but about 14 *square miles.* The *population* is differently estimated at from 700, to 2000 persons. The *inhabitants* are of a tawny colour, well formed, sagacious, and hospitable, yet thievish. The *surface* is mountainous; some of the *peaks are visible* at the distance of 45 miles.

9. SANDWICH ISLANDS.

Q. What is the *number* and *situation* of the *Sandwich Islands?*

A. They are 10 in number, and are the most northern of the islands in Oceanica.

* 1. Of these islands, Owhyhee, now called Hawaii, is the *largest,* and indeed the largest in Eastern Oceanica, being 97 miles *long,* and 78 *broad.* On this island Capt. Cook was *killed* by the natives.

2. The *climate* of these islands is similar to that of the West India islands, which lie in the same latitude, but more *temperate.* The *productions* are bread fruit, sugar cane, cocoanuts, sweet potatoes, &c. The *population* is recently estimated at 130,000.

3. On *Hawaii* are several *mountains* of great elevation. Mouna Roa, *2d class,* is the *highest.* Mouna Kea is *nearly as high.* Kerauea is a *frightful volcano,* whose upper crater is estimated at 7½ miles in circumference, and 1000 feet deep. The American missionaries *lately counted* 51 *craters* of different sizes, 21 of which were in a state of activity.

4. The *inhabitants* are mild, affectionate, and docile. In 1819, they *renounced idolatry* and burned their idols. Several of the natives have been educated at the Mission School at Cornwall, in Connecticut, and in company with missionaries from America, have returned to these islands. The *success* of these *missionaries* and heathen youth, in confirming the inhabitants in the Christian religion, has been as signal, as it was unexpected.

Questions.—Map of Oceanica.

What islands are included in Eastern Oceanica, or Polynesia? What two clusters bound it on the east? What two on the west? Where are the Pelew Islands? Which way from the Pelew Islands are the Ladrones? Which way the Carolines? Which way from N. Hebrides are the Friendly isles? What is the situation of the Navigators' Isles? What of the Society Isles? *In what latitude and longitude is Otaheite?* Where is Pitcairn's Island? In what direction from Pitcairn's island is Easter Island? What is the longitude *of this island?* Where are the Marquesas islands in respect to Otaheite? What islands are the most northern in the Pacific Ocean? What is the largest of these islands? *What is its latitude and longitude? How far from the tropic of Cancer?* What groups in Eastern Oceanica lie south of the equator? What north? *What great city in North America is in about the same latitude with the Sandwich islands?*

REVIEW.
I. Of Countries.

N. B. It would be. convenient for teachers to keep by them *written answers* to all the questions, in the following Review.

The pupil will make out his answers, as far as possible from *memory,* and the rest, partly from the *book,* and partly from the *Atlas.* It may be well for him to make out *written* answers, submit them to the teacher for correction, and then recite them from memory.

It will be observed that *Notes,* containing *General Remarks,* are added to some of the subjects; the teacher can examine the pupil in them, as he chooses.

Examination.

N. B. *Where the interrogation point is omitted in the following table, under any one of the subjects, the question respecting that subject and the country at the left hand of which the omission is found, is to be omitted.*

How is the Western continent *bounded?* North America? United States? New England? Here let the pupil be required to give the boundaries of each of the countries in the table one by one, as they are named by the teacher.

Describe the *face of the country* in the United States; New England; Maine, and of the other countries.

Describe the *soil* of New England; of Maine, &c.

Describe the *climate* of the W. Continent; of North America, &c.

What is the *extent,* that is, length, breadth, and square miles, of the Western Continent? of North America? &c.

What is the name and class of the *capital* of the United States? Of Maine? &c.

What is the name and class of the *chief town* in the United States? In Maine? &c.

What *universities and colleges* are in New England? In Maine? &c.

What are the *productions* of the United States? Of New England? &c.

What are the *manufactures* of New England? Of Massachusetts? &c.

What are the *exports* of the United States? Of New England? &c.

What can you say of *religion* in New England? In Maine? &c.

What is the *character* of the people of New England? Of the Southern States? &c.

What is the *government* of the United States? Of Mexico? Of Guatemala? &c.

	Pages	Face of the Country	Soil	Climate	Extent	Population	Capital	Chief town	Universities, Colleges	Productions	Manufactures	Exports	Religion	Character	Government
Western Continent	22	.	.	?	?	?	.	.	.	.	.	.	.	.	.
North America	23	.	.	?	?	?	.	.	.	.	.	.	.	.	.
United States	26	?	.	?		?	?	.	?			?		.	?
New England	30	?	?	?	.	.	.	.	?	?	?	?	?	?	.
Maine	32	?	?	.	?	?	?	?	?	?	.	?	?	.	.
New Hampshire	33	?	?	?	?	?	?	?	?	?	.	?	?	.	.
Vermont	35	?	?	?	?	?	?	?	?	?	.	?	?	.	.
Massachusetts	36	?	?	?	?	?	?	?	?	?	?	?	?	.	.
Rhode Island	39	?	?	?	?	?	?	?	?	?	?	?	?	.	.
Connecticut	40	?	?	?	?	?	?	?	?	.		?	?	.	.
Middle States	42	?	?	?	.	.	.	.	?	?	.		?	?	.
New York	44	?	?	?	?	?	?	?	?	?	?	?	?	.	.
New Jersey	47	?	?	?	?	?	?	?	?	?	?	?	?	.	.
Pennsylvania	48	?	?	?	?	?	?	?	?	?	.	?	?	.	.
Delaware	50	?	?	?	?	?	?	?	.	?	.	?	?	.	.
Southern States	51	?	?	?	.	.	.	.	?	?	.	?	?	?	.
Maryland	54	?	?	?	?	?	?	?	?	?	?	?	?	.	.
Virginia	55	?	?	?	?	?	?	?	?	?	.	?	?	.	.
District of Columbia	57	?	?	?	?	?	?	?	?	.	.	.	.	.	?
North Carolina	57	?	?	?	?	?	?	?	?	?	.	?	?	.	.
South Carolina	59	?	?	?	?	?	?	?	?	?	.	?	?	.	.
Georgia	60	?	?	?	?	?	?	?	?	?	.	?	?	.	.
Alabama	62	?	?	?	?	?	?	?	.	?	.	?	.	.	.
Mississippi	63	?	?	?	?	?	?	?	.	?	.	?	.	.	.
Louisiana	64	?	?	?	?	?	?	?	.	?	.	?	.	.	.
Western States	66	?	?	?	.	.	.	.	?	?	.	.	?	?	?
Tennessee	68	?	?	?	?	?	?	?	?	?	.	?	?	.	.
Kentucky	69	?	?	?	?	?	?	?	?	?	.	?	?	.	.
Ohio	70	?	?	?	?	?	?	?	?	?	.	?	?	.	.
Indiana	72	?	?	?	?	?	?	?	.	?	.	.	.	.	.
Illinois	73	?	?	?	?	?	?	?	.	?	.	?	.	.	.
Missouri	74	?	?	?	?	?	?	?	.	?	.	?	.	.	.
Michigan Territory	75	?	?	?	?	?	?	?	.	?	.	.	.	.	.
North Western Territory	76	?	?	?	?	?	.	.	.	?	.	.	.	.	.
Missouri Territory	77	?	?	.	?	.	.	.	.	.	.	.	.	.	.
Western Territory	78	.	.	.	.	.	.	?	.	.	.	.	.	.	.

	Pages	Face of the Country	Soil	Climate	Extent	Population	Capital	Chief town	Universities, Colleges	Productions	Manufactures	Exports	Religion	Character	Government
Arkansas Territory	79	?	?	?	.	?	?	?	.	.	.	.	.	.	.
Florida	79	?	?	?	?	?	?	?	.	?	.	.	?	?	?
Mexico	81	?	?	?	?	?	?	?	?	?	.	.	?	?	?
Guatemala	83	?	?	?	?	?	?	?	.	?	.	.	.	.	?
British America	85	.	.	.	.	.	.	.	.	.	.	.	.	?	?
Nova Scotia	86	.	?	?	?	?	?	?	.	.	?	.	.	?	?
New Brunswick	87	.	?	?	?	?	?	?	?	?	.	?	.	.	.
Upper Canada	88	?	?	?	?	?	?	?	.	?	.	?	.	.	.
Lower Canada	89	?	?	?	?	?	?	?	.	.	.	?	?	.	.
New Britain	90	.	?	?	.	.	.	.	.	.	.	.	.	?	.
Russian Settlements	91	?	.	.	?	.	.	.	.	.	.	.	.	.	.
Greenland	91	?	?	.	?	?	.	.	.	.	.	-	?	?	.
South America	95	.	.	?	?	?	.	.	.	?	.	.	?	?	.
Colombia	98	?	?	?	?	?	?	?	.	?	.	.	?	.	?
Guiana	100	?	?	?	?	?	?	?	.	?	.	.	?	.	.
Peru	102	?	?	?	?	?	?	?	.	?	.	.	.	.	?
Bolivia	103	?	?	?	?	?	?	?	.	?	.	.	?	?	?
Brazil	106	?	?	?	?	?	?	?	.	?	.	?	.	.	?
United Provinces	108	?	.	?	?	?	?	?	.	?	.	?	?	.	?
Chili	110	?	?	?	?	?	?	?	.	?	.	.	?	.	?
Patagonia	111	?	?	?	?	.	.	.	.	.	.	.	?	.	.
Eastern Continent	112	.	.	.	.	.	.	.	.	.	.	.	.	.	.
Europe	112	.	.	?	?	?	.	.	.	.	.	.	.	.	.
Km. Great Britain	116	.	.	.	?	?	.	.	.	.	.	.	.	.	?
England	117	?	?	?	?	?	?	?	?	?	?	?	?	?	.
Scotland	120	?	?	?	?	?	.	?	?	?	?	.	?	?	?
Ireland	122	?	?	?	?	?	.	?	?	?	?	?	?	?	?
France	124	?	?	?	?	?	?	?	?	?	?	?	?	?	?
Spain	127	?	?	?	?	?	?	?	?	?	?	?	?	?	?
Portugal	130	?	?	?	?	?	?	?	?	.	?	?	?	?	?
Italy	132	?	?	?	?	?	.	.	?	?	.	?	?	?	?
Lombardy	134	.	.	.	?	?	?	?	.	.	.	.	.	.	.
Sardinia	134	.	.	.	?	?	?	?	.	.	.	.	.	.	.
Modena, Lucca, and Parma	134	.	.	.	.	.	?	?	.	.	.	.	.	.	?
Tuscany	135	.	?	.	?	?	?	?	.	.	.	.	.	.	.

	Pages	Face of the Country	Soil	Climate	Extent	Population	Capital	Chief Town	Universities, Colleges	Productions	Manufactures	Exports	Religion	Character	Government
Chinese Empire	185	.													
China Proper	185	?	?	?	?	?	?	?	.	?	.	?	?	?	?
Corea	188	.	?	.	.		?	?	.	.	.	.	?	.	.
Chinese Tartary	189	?	?	?	?	?	.	.	.	.	.	.	.	?	.
Thibet	189	?	?	?	?	?	?	?	.	.	.	.	?	?	?
Japan	190	?	?	.	?	?	?	?	.	?	?	.	?	?	?
Africa	192	?	?	?	?	?	.	.	.	?	.	?	.	?	.
Northern Africa	195	.	.	.	.	.	.	.	.	.	.	.	.	?	.
Egypt	195	?	?	?	?	?	?	?	.	.	.	.	?	?	?
Barbary States	198	?	?	?	?	?	.	.	.	?	.	.	?	?	?
Morocco	199	.	.	.	.	?	?	.	.	.	.	.	.	.	.
Algiers	200	.	.	.	.	?	?	.	.	.	.	.	.	.	.
Tunis	200	.	.	.	.	?	?	.	.	.	.	.	.	.	.
Tripoli	200	.	.	.	.	?	?	.	.	.	.	.	.	?	.
Western Africa	201	?	?	?	.	.	.	.	.	?	.	.	.	?	.
Senegambia	202	.	.	.	.	.	.	.	.	.	.	.	.	.	.
Sierra Leone	203	.	.	.	.	?	?	.	.	.	.	.	.	.	.
Coast of Guinea	203	.	.	.	.	.	.	.	.	.	.	.	.	.	.
Coast of Congo	204	.	.	.	.	.	.	.	.	.	.	.	.	.	.
Southern Africa	205	.	.	.	.	.	.	.	.	.	.	.	.	.	.
Col. of C. G. Hope	205	?	?	?	?	?	?	?	.	.	.	.	.	?	.
Caffraria	207	.	.	.	.	?	?	.	.	.	.	.	.	?	.
Eastern Africa	208	?	?	?	.	.	.	.	.	.	.	.	.	.	.
Nubia	209	?	?	?	.	.	?	?	.	?	.	.	.	?	?
Sennaar	210	?	?	.	.	?	?	?	.	?	.	?	.	.	?
Abyssinia	210	?	?	?	?	?	?	?	.	?	?	?	?	?	?
Central Africa	213	.	.	.	.	.	.	.	.	.	?	.	?	?	
Nigritia	214	.	.	.	.	.	.	.	.	.	.	.	.	.	.
Oceanica	217	.	?	?	?	.	.	.	.	.	.	.	.	.	ᵣ

XX. Islands.

Let the pupil tell where and how each of the following Islands is *situated;* if it be an important island let the teacher turn to the page and ask its *extent, population,* &c. In case of a group of islands, as the West Indies, the teacher can turn to the page, and add such names of islands as he chooses.

Mt. Desert, Seguin, Deer, Long Is. p. 33. Nantucket, Martha's Vineyard, Elizabeth's Is. 37. Rhode Island, Block,

39, Long Is. Staten, Manhattan, p. 45, Sullivan's, James, Johns, Edisto, p. 59, Newfoundland, p. 86, St. Johns, p. 87, Breton, p. 90, West Indies, Cuba, Jamaica, Porto Rico, Martinique, St. Domingo, p. 92.

Terra del Fuego, Falkland Is. Juan Fernandez, p. 96, Margaritta, p. 100, Chiloe, p. 110. Staten, p. 111.

British Is. p. 116, Is. of Wight, Alderney, Guernsey, Jersey, Scilly Is., Is. of Man, Anglesea, p. 119, Hebrides Orkney, and Shetland Is. p. 122, Corsica, Rha, Belle Isle, Ushant, p. 126, Majorca, Ivica, Minorca, p. 128, Malta, p. 133, Sardinia, p. 134, Elba, p. 135, Sicily, p. 136, Candia, Negropont, Lemnos, Andros, Naxos, Paros, Mytylene, Scio, Patmos, Rhodes, Cyprus, p. 139, Ionian Isles, Cephalonia, Corfu, Zante, Cerigo, Ithica, Santa Mauro, Paxo, p. 140, Nova Zembla, Oesel, Dago, Aland, Spitzbergen, p. 155, Iceland, p. 158.

Ceylon, p. 180, Formosa, Hainan, Leeoo Keeoo, p. 188, Niphon, Ximo, p. 190.

Azores, Madeira, Cape Verde Is., St. Helena, Madagascar, p. 216, Comero, Bourbon, Mauritius, p. 217.

Sunda Is., Sumatra, Java, Borneo, p. 220, Philippine Is., Lucon, Mindanao, Spice Is., Celebes, Gilolo, 221, Ceram, Amboyna, Banda Is., Timorian Is., 222, New Holland, 223, Van Diemen's Land, New Guinea, New Britain, New Ireland, Louisiade, Solomon's Is., 224, New Caledonia, New Hebrides, New Zealand, 225, Pelew Is., Ladrones, Carolines, Friendly Is., 226, Navigators' Is., Society Is., Marquesas Is., 227, Easter, Sandwich Is., and Hawaii, 228.

XXX. Peninsulas.

Let the pupil describe the following peninsulas, by telling with *what country* they are *connected*—by what *waters* they are *nearly surrounded*—and, where practicable, in what *capes* they *terminate*.

Greenland, Labrador, Nova Scotia, Cape Cod, Michigan, Florida, Yucatan, Alaska, California, New Jersey.

South America.

Norway and Sweden, Spain and Portugal, Italy, Denmark, Turkey, Morea, Crimea.

Kamtschatka, Corea, Malaya, Hindoostan, Arabia, Turkey in Asia.

XY. Isthmuses.

Let the pupil tell *what countries* the following *isthmuses connect*. Darien, Suez, Corinth.

V. Capes.

Let the pupil describe the following *capes* by telling *where they are*, and into what *waters* they *extend*.

Farewell, Sable, Charles, Ann, Cod, Malabar, Montauk,

Sandy Hook, May, Henlopen, Henry, Hatteras, Lookout, Fear, Canaveral, Florida, Sable, Roman, Corrientes, St. Lucas, Mendocino, Blanco, Icy.

North, St. Roque, St. Maria, St. Antonio, St. Joseph, Blanco, Horn, Blanco.

North, Naze, Land's End, Cape Clear, La Hogue, Ortegal, Finisterre, St. Vincent, Trafalgar, Matapan.

Taymour, Lopatka, Comorin.

Bon, Nun, Bojador, Blanco, Verde, Roxo, Mesurado, Palmas, 2 Points, Coast, Formosa, Lopez, Gonsalvo, Negro, Voltas, Good Hope, Laguillas, St. Mary, Corrientes, Ambro, Delgado, Guardafui.

York, Flattery, Sandwich, Townsend, Sandy, Morton, Howe, Chatham.

VI. Plains and Deserts.

Let the pupil describe the following *Plains and Deserts*, by telling where they lie, their extent, &c.

Atacama, page 109, Beloochistan, 175 and 177, Sahara, 194, Barca, 201, Great Karoo, 206.

Notes.

1. The *countries* which *contain the most remarkable plains* are, Russia, Siberia, North and South America. *In Russia and Siberia these plains are called* steppes; *in North America*, prairies, or savannas; and *in S. America*, ilanos, or pampas.

2. The *plains or steppes of Russia and Siberia* are numerous and extensive. Some of them are nearly deserts; others are marshes, covered with long grass, and filled with salt lakes.

3. The *prairies of N. America lie principally* between the Alleghany and Rocky mountains. *Some of them* are low and marshy; others are elevated, and although fertile, are *called* barrens, because destitute of timber.

4. *The sea coast of the U. States*, extending from Long Island to the Gulf of Mexico, a distance of 1500 miles, is a plain from 50 to more than 100 miles wide. It abounds with barrens.

5. The *principal llanos, or pampas of South America*, are those of Venezuela, (Va-na-zwa'la) which extends 200 leagues along the River Oronoco; and the pampas of Buenos Ayres, (Bo-nos-Aires.)—p. 108.

6. *The countries which contain the most remarkable Deserts*, are Africa, Persia, Arabia, Chinese Tartary, and North America.

7. *The greatest desert on the globe*, is the Sahara of Africa.—p. 194. *Africa also contains the deserts* of Barca, Lybia, Nubia, Cimbebas, the extent of which is unknown.

8. The *principal desert of Arabia* is Tehama, extending along the coast from the Isthmus of Suez to the head of the Persian Gulf.—p. 173.

9. *Chinese Tartary contains the* desert of Cobi, or Shamo, 300 miles long, and 300 broad.—p. 184.

10. *N. America contains a great desert between the Platte river*

and the head of the Colorado and Sabine rivers. Length unknown; breadth supposed to be 500 or 600 miles.

UII. Mountains.

Let the pupil describe the following *Mountains,* i. e., tell where they are *situated,* and to what *class* they belong.

Alleghany, Rocky, 24, Green, White, 30, Monadnock, 34, Killington, Camel's Rump, Mansfield, 35, Wachusett, 37, Catskill, 44, Otter Peak, 55, Table Mountain, 59, Popocatepetl, Orizaba, 81.

Andes, 96, Chimborazo, Cotopaxi, 99.

Alps, Mont Blanc, Mont Rosa, Mt. St. Bernard, Mt. St. Gothard, Mt. Cenis, Pyrenees, Mt. Perdu, Carpathian, Appenines, Dofrafield, Ural, 113, 141, Ben Nevis, 121, Mt. Jura chain, Reculet, Dole, 125, Cantabrian range, Iberian range, Mts. of Castile, Mts. of Toledo, Sierra Morena, Sierra Nivada, 128, Mt. Etna, Vesuvius, 132.

Altay, Himmaleh, Dawalageri, Belur Tag, Mts. of China, Gauts, 164, 180, Mt. Taurus, Lebanon, Ararat, Hermon, Carmel, Tabor, 168, Caucasus, 171, Sinai, Horeb, 173

Atlas, Mts. of the Moon, Kong, 193, 198, Langekloff, Zwarte Berg, Snowy, Table, 206, Lupata, Abyssinia, 208, Geesh, Amid-Amid, Lamalmon, Gondar, 210, Peak of Teneriffe, 216, Ophic, 220, Tamboro, 222, Mouna Roa, Mouna Kea, 228

Notes.

1. The *longest range of mountains in the world,* is the American range, 11,500 miles.

2. *The longest range in Asia,* is the Altaian range, 5000 miles.

3. *The longest ranges in Africa,* are the Mountains of the Moon 2000 miles, and the Atlas range, 1500 miles.

4. *The longest range in Europe,* is the Ural range, 1500 miles.

5. *The Dofrafield range is* 1000 miles; *the Carpathian, 500; the Alleghany,* 900; *the Green* Mountains, 350; *the Alps and Appenines,* 700; *the Pyrenees, 200.*

6. *Mountains serve to* collect and condense the clouds and vapours, and thus supply the springs and streams which fertilize the earth—also to cool and equalize the temperature of the warmer regions—to arrest the progress of the winds; and to diversify and enrich the beauties of the landscape.

UIII. Volcanoes.

Let the pupil describe the following *Volcanoes*—that is, tell where they are *situated,* and other remarkable circumstances respecting them.

Popocatepetl, Orizaba, 81, Andes, 96, Cotopaxi, 99, Ætna, Vesuvius, 132, Hecla, 158, Geesh, 210, Tamboro, 222; Kerauea, 228.

Notes.

Volcanoes are burning mountains, with apertures called cra ters, out of which are thrown, with dreadful explosions, ashes, smoke, mud, fire, red hot stones, and lava.

2. *The number of volcanoes which have been discovered, is* about 200, one half of which are in America, and one quarter in Asia. Not less than 40 are said to be *continually burning, between Cape Horn and Cotopaxi.*

3. *The volcanoes of Europe and Asia* are generally on islands. *Those of America are* generally on the main land.

4. *The most celebrated volcanoes in the world are,* Mt. Etna, in Sicily, Vesuvius in Italy, Hecla in Iceland, and Cotopaxi in South America.

5. *Volcanoes are ascribed* to fires in the bowels of the earth, caused in some way not well understood. *Some attribute these fires to* the inflammation of bitumen, pit coal, fossil wood, &c.; others suppose them to arise from the inflammation of pyrites, a metallic substance formed of iron and sulphur.

6. *Volcanoes serve as* the outlets of these fires, which, if confined, would produce the most desolating earthquakes.

7. *Earthquakes are caused,* it is supposed, *by* some volcano which usually explodes at the same time.

8. *Countries most subject to earthquakes* are situated in volcanic regions. Hence earthquakes are frequent in the countries bordering on the Mediterranean sea, and in those which surround the Carribbean sea; in the Gulf of Mexico, in South America, particularly in Peru.

9. *The principal volcanic chain may be represented by* a line drawn in the following manner: begin at Terra del Fu'ego, in South America, through the whole American range (Andes, Cordilleras, and Rocky mountains,) to Behring's strait; thence to Kamtschatka; thence by a line drawn through Japan, Formosa, Java, Sumatra, New Amsterdam, and St. Paul's (islands in the Indian Ocean,) Gebel Tar, (in the Red Sea,) thence through Syria, Greece, Italy, Germany, southern part of France, and even through England and Scotland to Iceland. The middle of the Atlantic Ocean conceals another volcanic focus, of which the Azores and Canary Islands have felt the influence. No less than 42 volcanoes, either active or dormant, are found among the Azores.

IX. Caves.

Let the pupil describe the following *Caves*—that is, tell where they *are,* and the other remarkable circumstances about them.

Madison's, Wier's, Nickojack, Cave in the Alleghany Ridge, of Frederickshall, Fingal's, of Antiparos, Peak Cavern, and Grotto del Cane.

Notes.

1. *Caves are most frequent in* countries abounding with limestone.

2. *Some of the most remarkable caves in the United States* are, 1. Madison's, in Rockingham County, Virginia, extending 300 feet into the earth, and adorned with beautiful incrustations of stalactites. 2. Wier's, in the same county, extending 800 yards, but extremely irregular in its course and size. 3. Nickojack, in Georgia, is 50 feet

high, and 100 wide. It has been explored to the distance of 3 miles. A remarkable circumstance respecting this cave is, that a stream of considerable size runs through it, which, at a distance of 3 miles, is broken by a water fall. 4. In the Alleghany Ridge, in Virginia, is a blowing cave, from which wind constantly issues. It is 100 feet in diameter, and the current of air is so strong as to keep the weeds prostrate to the distance of 60 feet from its mouth. 5. Besides these caves, several others of great dimensions are to be found in Kentucky, Tennessee, and other places.

3. *Some of the most remarkable caves in Europe are*, 1. The cave of Frederickshall, in Norway, which is supposed to be not less than 11,000 feet in depth. 2. Fingal's Cave, which is doubtless the most magnificent of the known caverns in the world. Thousands of majestic columns of basalt, or marble, support a lofty roof, under which the sea rolls its waves, while the vastness of the entrance allows the light of day to penetrate the various recesses of the cave. Its length is 370 feet, and its height, at the entrance, 117 feet. 3. The Grotto of Antiparos has long been celebrated for its beauty and magnificence. The principal grotto is at the depth of 1500 feet, and is 360 feet long, and 340 wide, and 180 high. The roof is adorned with crystals called stalactites, some of which are 20 feet long, intermingled with massy columns, which descend to the floor. 4. The Peak Cavern of Derbyshire, is a celebrated curiosity of a similar kind. It is nearly half a mile in length, and 600 feet deep. 5. The Grotto del Cane, near Naples, in Italy, sends forth a constant stream of carobonic acid gas, or fixed air, which, however, being heavier than common air, rises to a small height only, from the ground. From this circumstance, a man may walk into the cave upright without injury, while an animal of small height meets the current, and immediately dies.

X. Oceans.

Let the pupil describe the following *Oceans*, by telling *where they are*, and if the teacher chooses, the *extent* of each, according to pages 15, 16.

Northern, Southern, Pacific, Indian, and Atlantic.

Notes.

1. *The level of the ocean* is the same in every part of the world, excepting in gulfs and inland seas. The Red Sea is thought to be 30 feet higher than the Mediterranean. The waters of the gulf of Mexico are 20 feet higher than those of the Pacific.

2. *The saltness of the ocean* appears to be less near the mouths of rivers, and gradually diminishes in going towards the poles.

3. *As to the origin of the saltness of the ocean*, much diversity of opinion exists ; the most rational opinion is, that it was created salt.

4. *The depth of the ocean* is extremely various, its bed being diversified like the land, with hills, valleys, and mountains. *The greatest depth yet sounded*, is said to be only 7200 feet.

5. *The effect of the waves of the sea does not extend* beyond a certain depth. The divers assure us that in the greatest tempests calm water may be found at the depth of 90 feet.

6. *In the southern hemisphere, permanent ice begins* in about lat. 72°. *In the northern,* in about lat. 80°, although floating ice may be found at all seasons as low as lat. 70°.

7. *Icebergs are* lofty masses of ice formed in the polar regions, and are wafted by the wind, waves, and currents, sometimes as low as lat. 40°. Some of these icebergs are half a league in *length,* and 600 feet *high.*

8. *The luminous appearance of the water* which a vessel throws on either side, in her progress on the ocean in the night, *is attributed to* myriads of small animals, which float upon the surface of the sea, and which, when agitated, have the power of emitting light.

9. The incessant agitation of the ocean united to its saltness, *prevents its waters from corrupting.*

XI. Seas.

Let the pupil describe the following *Seas* by telling *what countries border upon them,* and *with what waters they are connected.*

Carribbean.

White, North, Baltic, Irish, Mediterranean, Marmora, Black, Azof.

Red, Arabian, China, Yellow, Japan, Ochotsk, Kamtschatka, Anadir.

XII. Bays.

Let the pupil describe the following *Bays* by telling *what countries border upon them,* and *with what waters they are connected.*

Baffin's, Hudson's, James, Fundy, Passamaquoddy, Machias, Penobscot, Casco, Massachusetts, Buzzard's, Naragansett, Delaware, Chesapeake, Campeachy, Honduras, Bristol.

All Saints.

Cardigan, Donegal, Galway, Biscay.

Bengal.

Walwich, Table, False, Angola, Natal, Saldanha.

Botany Bay.

XIII. Gulfs.

Let the pupil describe the following *Gulfs* by telling *what countries border upon them,* and *with what waters they are connected.*

St. Lawrence, Mexico, Amatique, Darien, California.

Panama, Guayaquil, St. George.

Bothnia, Finland, Riga, Lyons, Genoa, Naples, Tarento, Venice, Salonica.

Persian, Ormus, Siam, Tonquin, Corea, Obi.

Guinea.

XIV. Straits.

Let the pupil describe the following *Straits* and *Channels,* by telling *what waters they connect,* and *what lands they separate.*

Davis's, Hudson's, Belle Isle, Michillimackinac, Behring's.

Magellan.

Skager Rack, Cattegat, Dover, Gibraltar, Bonifacio, Messina, Dardanelles, Constantinople, English, St. George, and North.

Babelmandel, Ormus, Malacca.

Mozambique.

Macassar, between Borneo and Celebes, Sunda, Malacca, Endeavour, Bass's, Dampier's, and Cook's.

XV. Sounds.

Let the pupil describe the following *Sounds* by telling *what countries border upon them* and *with what waters they are connected.*

Long Island, Albemarle, Pamlico, Prince William's, Queen Charlotte's, and Nootka.

XVI. Lakes.

Let the pupil describe the following *Lakes*, by telling *where they are situated*, and, if the teacher chooses, their *Length and Breadth.*

	Length	Breadth		Length	Breadth
Superior,	490	100	Champlain,	128	15
Huron,	220	90	Chapala,	54	15
Michigan,	300	50	Pontchartrain,	35	25
Winnipeg,	250	50	Winnipiseogee,	23	10
Erie,	230	45	Memphremagog,	35	3
Slave,	270	50	George,	36	2
Ontario,	180	40	Seneca,	35	2 to 4
Nicaragua,	120	41	Cayuga,	38	1 to 4
L. of the woods,	70	40	St. Clair,	90	in cir.
Maracaybo,	200	70	Titicaca,	240	in cir.
Ladoga,	140	75	Wetter,	65	16
Onega,	130	70	Geneva,	50	10
Wenner,	80	30	Constance,	40	10
Malar,	80	20	Neagh,	15	8
Lomond,	30	9			
Caspian Sea,	640	260	Baikal,	360	20 to 50
Aral Sea,	250	120	Asphaltites, or }		
Maravi,	300	30	Dead Sea, }	65	15
Dembea,	450 in cir.				

Notes.

1. *The great lakes of North America discharge their waters* through the St. Lawrence; *the Slave Lake through* the Mackenzie River; *Nicaragua through* the St. Juan; *Champlain through the* Sorelle into the St. Lawrence; *Chapala through* the Santiago. *Pontchartrain communicates* with the Gulf of Mexico; *Winnipiseogee runs* into the Merrimack; *Memphremagog through* St. Francis into St. Lawrence; *George communicates* by an outlet with Lake Champlain; and *Seneca and Cayuga pass* through Oneida River into Lake Ontario.

2. *Maracaybo lake empties* into the Carribbean Sea.

3. *Lake Ladoga passes through* the Neva, into the Baltic; *Onega through* the river Svir into Lake Ladoga; *Wenner through the* Gotha into the Cattegat; *Malar communicates* with the Baltic at Stockholm; *Wetter passes through* the Motala into the Baltic; *Geneva through* the Rhine; *Constance through* the Rhone; *Lough Neagh through* the river Bann; *and Loch Lomond through* a short outlet into the Clyde.

4. *Lake Baikal discharges its waters through* the Angara; the Caspian Sea, Sea of Aral, and most of the lakes in Asia, Africa, and South America, *have no visible outlet.*

5. Those which have an outlet are *fresh;* those which have no outlet are *generally salt.*

6. *The navigation of inland waters is generally more dangerous than that of the ocean, because* the winds on these waters are rendered more unsteady by the mountains which interrupt and vary them.

XVIII. Rivers.

Let the pupil describe the following *Rivers,* i. e. tell their *class,* where they *rise,* which way they *flow,* through what countries they *pass,* and where they *empty.*

Mackenzie's, Nelson's, St. Lawrence, Mississippi, Missouri, Rio del Norte, Colorado, Columbia, p. 24; Connecticut, 30; Penobscot, Kennebeck, Androscoggin, Saco, Piscataqua, 33; Pawtucket, Providence, Pawtuxet, 39; Housatonick, 41; Hudson, Delaware, Susquehanna, 43; Mohawk, Genesee, 44; Raritan, Passaick, 47; Brandywine, Christiana, 51; Potomac, Elk, Chester, Choptank, Nanticoke, Wicomico, Pocomoke, 54; James, Rappahannock, Great and Little Kenhawa, 56; Roanoke, Neuse, Pamlico, Cape Fear, 58; Santee, Pedee, 60; Savannah, Alatamaha, 61; Chatahoochee, Mobile, Alabama, Tombeckbee, Black Warrior, 62; Yazoo, Black, Pearl, Pascagoula, 64; Red, Washita, 65; Cumberland, Tennessee, 68; Kentucky, Licking, Green, 70; Ohio, Muskingum, Hockhocking, Scioto, Great Miami, Maumee, Sandusky, Cuyahoga, 71; Wabash, Tippecanoe, White, 72; Illinois, Kaskaskia, Rock, 73; Yellowstone, Platte, Kansas, 77; St. Johns, Appalachicola, 80; St. Juan, 84; and St. Johns, 88.

Amazon, La Plata, Oronoco, Parana, Uruguay, Madeira, Tocantins, Magdalena, 96; Meta, Cauca, 99; Ucayle, 102; Pilcomayo, 104; St. Francis, 106; Xingu, Tapajos, 107; and Vermejo, 109.

Volga, Don, Dnieper, Dneister, Danube, Rhine, Rhone, Elbe, 115; Severn, Thames, Humber, Mersey, 117; Tweed, Forth, Tay, Clyde, 121; Shannon, Barrow, Liffy, Boyne, 123; Garonne, Loire, Seine, 125; Tagus, Guadiana, Guadalquiver, Ebro, Douro, Minho, 129; Po, 132; Save, Pruth,

Notes.

1. The *water of rivers is chiefly supplied by* springs, rains, and the melting of snow and ice.

2. The *magnitude of rivers generally depends upon* the extent of country through which they pass.

3. The *quantity of water discharged, generally depends upon* the elevation and extent of the mountains which border on the river.

4. The *rapidity of a river depends upon* the slope, or descent of the country through which it passes. When an impulse, however, has once been given to the mass of water, the pressure will keep it in motion, even if there were no declivity. Many great rivers, in fact, flow with an almost imperceptible declivity. The Amazon has only ten feet and a half of declivity, upon 200 leagues in extent of country, or 1-27th of an inch, for every 1000 feet. The Seine, in France, between Valvins and Serves, has only one foot declivity, out of 6600.

5. Those rivers which pass through level or alluvial regions are *furthest navigable ;* those which pass through mountainous regions are *least navigable.*

6. Those rivers, whose course is the straitest, *are most rapid,* and those most winding, *the least rapid.*

7. Those rivers which pass through mountainous regions, have the *highest banks ;* those which pass through level regions, *the lowest.*

8. The St. Lawrence is said to be the *only river not affected occasionally by floods.* Most rivers, especially those of the Torrid Zone, are *greatly affected by floods.* The Nile, Ganges, and Mississippi, sometimes rise 30 feet; the Ohio 40 or 50; The Oronoco from 70 to 120. The floods of the Oronoco, Amazon, and Ganges, often cover the country for hundreds of miles in extent. In 1822 the flood of the Ganges was supposed to have destroyed from 50,000 to 100,000 persons.

9. *Rivers are affected by the tide,* at different distances, according to the absence of obstructions of the channels and currents. The tide affects the Amazon 400 miles from its mouth; the Thames 70; the Connecticut 50; the Hudson 160; the Potomac 200.

10. *Those rivers* only which have large mouths, like the Loire, the Elbe, or the Plata, *mingle their waters peacefully with the ocean.*

11. Some rivers, however, on entering the ocean, especially when they meet the tide, produce an elevated ridge of waters which overturns boats, inundates the banks, and causes great destruction.

The most sublime phenomenon of this kind presents itself in the river Amazon. Twice a day it pours out its imprisoned waters into the ocean. A liquid mountain is thus raised to the height of 180 feet; it frequently meets the flowing tide of the sea, and the shock of these two bodies of water is so dreadful that it makes all the neighbouring islands tremble; the fishermen and navigators fly from it with the greatest consternation.

XVIII. Falls.

Let the pupil describe the following *Falls.*

Niagara; Montmorenci; Mississippi; Missouri; Passaic; Cohoes; Genesee; Bellows Falls; Tequendama; Fyers; Lattin; Terni; Tivola; and the cataracts of the Rhine and Nile.

Notes.

1. *Under the term falls are included* rapids, cataracts, and cascades. *Rapids* are that part of a river where the water is rapid over a moderate descent. A *cataract* is a fall of water over a precipice. A *cascade* is a succession of small cataracts.

2. *Falls are usually found* in those parts which are rugged and mountainous.

3. *The most distinguished cataract known in the world,* is the cataract or Falls of Niagara, in North America.—p. 46.

4. *Some of the other principal falls in North America,* are the Falls of the Montmorenci, 9 miles below Quebec; they are 220 feet in height; but the body of water is small, and the breadth of the river only 50 feet.

5. *The falls of the Mississippi,* above its junction with the Ohio, are 40 feet in height, and 700 feet in width. They are not remarkable, however, for their grandeur.

6. *The falls of the Missouri,* 500 miles from its sources, consist of three principal cataracts, the first 87 feet in height; the second 47; and the third 27. The river is 1000 feet broad. These falls are said to be but little short of those of Niagara in grandeur.

7. *The falls of the Passaic,* 15 miles from Newark, N. Jersey, are much celebrated. The river is 120 feet broad, and falls in one entire sheet into a chasm, 70 feet deep, and 12 wide.

8. The *falls of the Mohawk,* called the Cohoes, (Cahoze') are two miles from its mouth. The river descends in one sheet nearly 70 feet.

9. The *falls of the Genesee,* below Rochester, are 96 feet perpendicular descent, and when the river is high, are exceedingly grand.

10. *The falls of the Connecticut,* near Walpole, are called Bellows Falls. The river when low, is compressed into a rocky passage 16 feet in width, and rushes down with immense force and tumultuous roar.

11. *The highest fall or cataract in all America,* is that of Tequendama, (Ta-kuen-da'ma) on the river Bogota', 15 miles southwest of Santa Fe. The river just above the falls, is compressed from 140 to 35 feet in width, and rushes down a perpendicular rock at two bounds, to the astonishing depth of 600 feet.

12. *The most remarkable falls in Europe*, are the Falls of Fyen a river which flows into Loch Ness, in Scotland. These falls are two, one 70 feet, and the other 207 feet in height.

13. *The falls of Lattin*, in Swedish Lapland, have recently been discovered. The river at the falls is said to extend 2500 feet in width, and the perpendicular descent of water to be 400 feet.

14. *The falls of Terni*, in Italy, are in the river Eveline. 45 miles from Rome. Their height is 300 feet over a precipice of marble rock.

15. *The falls of Tivola*, in Italy, are in the river Anio, a branch of the Tiber, 18 miles N. E. of Rome. These falls are nearly 100 feet in height.

16. *The cataract of the Rhine*, near Schaffhausen, in Switzerland, is much celebrated for its grandeur. The river is 450 feet broad, the height of the fall 70 feet.

17. *The most celebrated falls in Africa*, are the falls or cataracts of the Nile. The first of these is near the city of Syene, and the other at some distance above. Although the highest does not exceed 40 feet, yet they are said to be very sublime, and to derive additional grandeur from the wildness and desolation of the surrounding scenery.

XIX. Springs.

Let the pupil describe the following *Springs*; that is, tell *where they are, of what kind they are*, &c.

Geysers; of Bath, Bristol, Buxton; of Barege, Vichy; Aix la Chapelle; Bath in Virginia, New Lebanon; Harrowgate, Bedford, York, Stafford; Spa, Pyrmont, Tunbridge, Wells, Brighton, Ballston; Cheltenham, Epsom, Seltzers, Saratoga.

Notes.

1. The *principal classes of springs are*, common, intermittent, warm. or hot, sulphureous, chalybeate, saline, aerated, or sparkling. *Common springs* are those which are marked with no peculiarity; *intermittent springs* rise and fall alternately; *warm* or *hot*, have a more than ordinary temperature; *sulphureous*, contain sulphur; *chalybeate* contain iron; *saline* contain various kinds of salts; *aerated*, or *sparkling*, contain carbonic acid gas, or fixed air, which causes them to sparkle. Many springs have several of these qualities united. Those which contain mineral qualities are termed mineral springs.

2. The *water of common springs is supplied principally by* the filtering of rain, dew, and melted water into the earth. *Water filters down* to very great *depths*. In the coal mines of Auvergne, it has been seen to penetrate as far as 250 feet. In Misnia, a town and district of Saxony, rain water has been observed to distil in drops from the roof of a mine, 1600 feet deep. The water of other springs, excepting the intermittent, probably pass over the mineral substances with which they are impregnated.

3. The *most noted springs of the intermittent kind*, are the Geysers, or spouting springs of Iceland.—p. 158. *Spouting springs are accounted for by some*, by supposing that their waters in some way becoming confined, burst forth in consequence of the pressure, just as water spouting fountains, which are formed in some gardens.

Others suppose the waters are sent forth by vapours formed sud denly by some means in the earth.

4. *The most remarkable hot springs of England*, are those of Bath, Bristol, and Buxton. The Bath springs are from 93 to 117 degrees ot temperature. They are found to be very efficacious in bilious com plaints, in rheumatic and gouty affections, palsies, &c. The hot wells of Bristol are only about 74 degrees of temperature; they contain considerable carbonic acid gas, and are used with much benefit in cases of dyspepsia and pulmonary consumption. The springs ot Buxton are 92 degrees; they are used externally in cases of chronic rheumatism, and internally as a remedy for defective digestion.

5. *The most celebrated hot springs in France*, are those of Barege, and Vichy. The springs of Barege are situated about half way between the Mediterranean and the Bay of Biscay. They are saline and sulphureous, and are from 80 to 123 degress of temperature. They are used in rheumatic and gouty affections, eruptions, colics, jaundice, &c. The waters of Vichy, in the department of Allier, are of the temperature of 120 degrees, and are chalybeate and alkaline in their properties. They are used in bilious complaints, diarrhœas, and rheumatisms.

6. *The most celebrated hot springs in Germany*, are those of Aix la Chapelle, near the borders of the Netherlands, in the Prussian Dominions. These waters are highly sulphureous, and are of the temperature of 143 degrees. They are the most celebrated mineral springs in Europe, being the usual resort of those, who are diseased from high indulgence in the luxuries of the table.

7. *The most celebrated warm and hot springs in the United States*, are, the one of Bath, in Virginia, and of New Lebanon, in New York. The warm spring of Bath discharges water sufficient to turn a mill, and keep the water in its basin, 30 feet in diameter at the temperature of 96 degrees. The hot spring is 6 miles distant and has a temperature of 106 to 108 degrees. The spring at New Lebanon has only a moderate degree of heat; but is found a valuable remedy for many diseases.

8. *The most celebrated sulphureous springs* are those of Harrogate, in England; and in the United States, those of Bedford and York, in Pennsylvania, and Stafford in Connecticut.

9. *The most celebrated chalybeate springs*, are those of Spa, in the Netherlands; Pyrmont, in Hanover; Tunbridge Wells, and Brighton, in England; Ballston, in New York, in the United States. These springs contain more or less carbonic acid gas, and are more or less impregnated with iron. They are used in cases of irregular digestion, general debility, &c.

10. *The most remarkable saline springs*, are those of Cheltenham and Epsom, in England; Selters or Seltzers, in Germany; Salina and Saratoga in the United States.

XX. Canals.

Let the pupil describe the following *Canals:*
The Erie Canal in N. York, page 46; the canals of England, p. 118; the Languedoc Canal in France, p. 125; and
21*

the Imperial Canal in China, p. 187; Canal of the Centre Burgundy, Caledonia, Grand Irish, Royal, Middlesex, Merrimack, Santee, Champlain, Ohio, and Farmington.

Notes.

1. A *Canal* is an artificial excavation of the earth, formed to contain water, chiefly for internal navigation, but sometimes for watering the lands, for mills, and other purposes.

2. The *embankments of a Canal* are ridges of earth, raised to prevent the water from overflowing; *aqueducts* are structures generally composed of stone, or brick, and thrown across valleys and rivers, to contain the water upon which the boats are to pass; *tunnels* are subterranean passages cut through mountains and hills, which cannot be cut down; *locks* are tight reservoirs, built on declivities, and closed by flood gates, at both ends, and are alternately filled and emptied to enable boats to ascend, or descend, with safety.

3. We may notice *several canals in Europe*; the canal of the Centre, in France, which unites the Saone to the Loire, 71 miles long; and the Canal of Burgundy, which unites the Saone and the Seine, about as long as that of Languedoc; the Caledonian canal in Scotland, which connects the Murray Frith with the Atlantic This canal is 59 miles in length, of which 37 is in natural waters. and is navigable for frigates of 32 guns. The Grand Irish canal connecting the Shannon and the Liffy. This canal is 83 miles long. Another canal called the Royal Canal, lies north of it, and runs nearly parallel with it from Dublin to the Shannon.

4. The *first canal of magnitude ever completed in the United States,* was the Middlesex canal in Massachusetts, connecting the river Merrimack with Boston harbour. The Merrimack canal was begun in 1790. It is 31 miles long, and admits boats of 14 tons. The *second* canal executed in the United States, was the Santee, in South Carolina, connecting the Santee and Cooper's river, 22 miles long. The *third* was the Champlain canal, connecting Champlain and the Hudson, also 22 miles long. Several *other canals of considerable length have been commenced,* and will be completed in the course of a short time. The principal of these are the Ohio canal, connecting the waters of Lake Erie with those of the river Ohio; and the Farmington canal, extending from the state of Massachusetts, to the sound at New Haven.

XXX. Latitudes and Longitudes.

What *Capitals* lie between the equator and 10 degrees of N. latitude?

N. B. Before the pupil answers, let him examine all the maps, and let his answer embrace all the capitals he finds between the equator and 10 degrees of north latitude. Let him adopt a similar rule in answering the following questions.

What Capitals lie between 10 and 20 degrees of N. latitude? between 20 and 30? between 30 and 40? between 40 and 50? between 50 and 60? between 60 and 70? &c.

What Capitals lie between the equator and 10 degrees of S. latitude? between 10 and 20? &c.

What Capitals lie between the meridian of London, and 10 degrees of east longitude? between 10 and 20? and so on to 180 degrees.

What Capitals lie between the meridian of London, and 10 degrees of west longitude? between 10 and 20? and so on to 180 degrees.

What *Islands* lie between the equator and 10 degrees of N latitude?

N. B. In some cases, islands do not lie wholly within the degrees of latitude embraced in the question; let the pupil in his answer therefore, have regard to the *middle*, or *central point* of the island as nearly as his eye can direct him.

What islands lie between 10 and 20 degrees of N. latitude?

Here the teacher can put the questions, as for capitals.

This examination, as to latitudes and longitudes may be carried to a great extent, embracing *capes, peninsulas, mountains, lakes, seas, bays, mouths of rivers,* &c. &c. It may be a useful exercise to require the pupil to make out *written answers* for them, if not, to recite them from memory.

XXXX. Vegetables.

In what latitudes do *wheat, flax,* and the *oak* flourish? *rye, oats, barley?* and so of the others.

N. B. The climate of Europe in northern latitudes is warmer than in corresponding latitudes in America. The difference may be equal to 7 or 8 degrees; that is, the temperature of 45 N. lat. in America, may be considered about equal to 52 or 53 in Europe. The northern limit of a vegetable or animal, therefore, may be considered 7 or 8 degrees less in America than in Europe, provided that limit is north of 30 N. latitude. The following limits of vegetables for north latitude are calculated for the *eastern* continent; consequently, 7 or 8 degrees are to be subtracted from these limits to make them answer for the *western* continent.

These notes on vegetables and animals are derived from Mr. Woodbridge's Physical Chart.

Notes.

1. Wheat, flax, and the oak, flourish between 24 and 65 degrees of north latitude, and between 24 and 48 of south latitude. Rye, oats, barley, between 25 and 66 N. lat. and 24 and 48 S. Maize, tobacco, between 54 N. and 44 S. The potato, between 68 N. and 54 S. Pepper, ginger, between 22 N. and 20 S.

The plum and cherry, between 33 and 55 N. and 30 and 46 S. The plantain, banana, yam, pine apple, cocoanut, and tamarind, between 21 N. and 20 S. The apple and pear, between 32 and 55 N. and 30 and 46 S. The orange and lemon, between 38 N. and 31 S. The peach, between 30 and 48 N. and 25 and 46 S. Berries, between 46 and 70 N. Indigo, rice, cotton, the fig and olive, between 44 N. and 33 S. The wine-grape, between 36 and 50 N. and 30 and 47 S. Coffee, sugar-cane, palm tree, between 36 N. and 32 S. The pine, and fir, between 30 and 70 N. Birch, between 38 and 70 N. Sugar-maple, between 38 and 60 N. Cacao, or chocolate nut, between 22 and 30 S., and Matte, a plant resembling tea, between 18 and 33 S. The three last on the western continent only.

2. The *earth produces* between 50 and 60,000 *kinds of plants.* The torrid *zone contains the greatest number of kinds; the earth diminishes* as the latitude increases N. or S.

XXXIII. Animals.

In what latitudes are the *dog, fox, cat, deer, rabbit, rat, and mouse* found? and so of the others.

Notes.

1. The dog, fox, cat, deer, rabbit, rat, and mouse, are found between 70 N. and 54 S. lat. The ox, horse, sheep, goat, hog, and wild boar, between 63 N. and 53 S. The ass, and mule, between 58 N. and 47 S. The reindeer, between 64 and 70 N. The wolf and bear, between 70 N. and 32 S. The ermine and sable, between 55 and 70 N. The beaver, marten, otter, between 37 and 70 N. The walrus, between 63 and 70 N. The alligator and crocodile, between 32 and 35 N. The buffalo or bison, is found west of the U. S., between 16 and 63 N. lat. The catamount, between the equator and 60 N. The cougar, between 41 S. and 60 N. The jaguar, between 40 S. and 8 N. The llama, vicuna, and guanaco, valuable beasts of burden, between 50 S. and 6 N.; these are peculiar to America.

2. The camel is found in Asia, Africa, and Russia, between 5 and 46 N. The lion in Asia and Africa, between 32 S. and 40 N. The elephant and rhinoceros, in Asia and Africa, between 24 S. and 30 N. The tiger, in Asia, between 12 S. and 40 N.

3. The leopard, hyena, hippopotamus, cameleopard, zebra, and ostrich, are found in the middle regions of Africa.

XXXV. Minerals.

What *countries produce gold in the greatest abundance?* What other countries produce gold? The pupil will be able to answer from the subjoined notes.

Let similar questions be put in respect to the following minerals, which the pupil will be able to answer from the book, and from the subjoined notes: viz. Silver, iron, copper, lead, tin, quicksilver coal, and salt.

Notes.

1. Those *portions of the earth* which are barren of vegetation are *richest in minerals.* Mineral substances are distributed as they are necessary to man; that is, the most useful abound most. *Metals are sometimes found in* a pure state, but *usually* in a stony substance, called ore. Some are found upon or near the surface of the earth, others are obtained at the depth of thousands of feet.

2. *Gold is usually found* in grains or gold dust in a pure state at the foot of long ranges of mountains, from which it is washed down by the rains and rivers. *The countries which produce gold in the greatest abundance,* are Brazil, Mexico, East and West Africa, and the islands of Borneo, Sumatra, and Celebes.

The only gold mines of importance in Europe, are at Kremnitz, in Austria, which produce more than all the rest of Europe. Mexico and South America *produce more gold than all the world besides.* Some gold has within a few years, been found in North Carolina.

3. The *silver mines* of Mexico and S. America are the *richest in the world.* The first silver mines in South America, were discovered at Potosi, by an Indian, who tore up a bush in ascending the mountains, under which he discovered a mass of silver. From these

mines in the course of three centuries, it is estimated that ;16,000,000 lbs. of pure silver have been obtained. The quantity of silver found in other parts of the world is small.

4. *Of all the metals* iron is the *most useful and the most generally diffused. The most productive mines occur in* Great Britain and France, which countries produce about twice as much iron, as all the rest of the world. It is, however, found in almost every country.

5. Great Britain *produces more copper and lead than all the rest of Europe.* The *principal copper mines* are those of Cornwall. Copper is also found in Norway, Sweden, Austria, and many other parts of the world.

Lead is found more or less in all countries. Mines, some of which are very productive, occur in France, Germany, Austria, and Spain. The lead mines of Missouri, in the United States, are very rich. The ore is found abundantly within two feet of the surface, in detached masses, weighing from 1 to 1800 lbs. The annual produce is estimated at 3,000,000 of pounds.

6. *Tin is chiefly found* in Cornwall in England. It is found in much smaller quantities in Saxony and Spain. The island of Banca, near Sumatra, is said to be almost entirely composed of it.

7. *The only mines of quicksilver of importance,* are those of Almaden, in Spain ; Idria, in Austria ; and Guanca Vilica, in Peru.

8. England *is the most remarkable for coal mines ;* the principal of which are at Newcastle, on the northeastern side of the island, and at White Haven, on the western. *Coal is also abundant* in various parts of France, Germany, China, and the United States. Large mines of coal are now wrought near Richmond, Virginia ; near the heads of the Schuylkill and Lehigh, Pennsylvania ; also near Pittsburg in the same state. A mine has also been lately re-opened in Rhode Island.

9. *Salt is chiefly derived from sea water,* which is extensively evaporated at Cape Verde and Turks Islands ; from beds of rock salt which occur in many parts of the world, particularly in England, Germany, Hungary, and Poland ; and from brine springs, the richest of which are at Salina, in New York.

XXV. Races of Men.

How many *varieties* does the human race exhibit ? What can you say of the *European race? The Asiatic ? American ? Malay? Ethiopian ? Hottentot?*

Notes.

The human family, doubtless, as the scriptures inform us, originally *sprung from* a single pair. In the lapse of time, however, owing to a difference of climate, manners, and modes of living, and other causes, probably beyond the reach of our investigations, *they exhibit six varieties* in respect to complexion, form, and character, viz. : The European or white race ; the Asiatic or tawny race ; the American or copper-coloured race ; the Malays, or dark brown race ; the Ethiopian, or black race ; and the Hottentot, or blackish race.

The *European race* has regular features, and in temperate climates a fair complexion ; but in warm climates their complexion is swarthy. *This race includes* all European nations and their descendants, except the Laplanders ; also the Circassians, Georgians, Arabians, Turks, Persians, and Hindoos.

2. The *Asiatic race* has a tawny, brown, or olive complexion, coarse, straight, black hair, small eyes, and high cheek bones. is generally stouter than the Europeans, but not so well formed. *This race includes* the Tartars, Chinese, and other inhabitants of the eastern and southern parts of Asia, excepting the Malays, and the Laplanders.

3. The *American race* has a copper colour, resembling that of rusty iron, or cinnamon, coarse, straight, black hair, high cheek bones, and sunken eyes. Their forehead is short, nose and countenance broad, nostrils open, lips thick ; *this race includes* all the American tribes of Indians.

4. The *Malay race* is of a nut brown colour, with black eyes, soft black hair, broad nose and mouth, and upper jaw projecting. *this race includes* the inhabitants of Malaya, Ceylon, Asiatic Islands, and Polynesia.

5. The *Ethiopian race* has black eyes, black, woolly hair, flat noses, thick lips, and a projecting upper jaw. They are generally short, stout, and ill formed ; *this race includes* the negroes of Africa.

6. The *Hottentot race* is a yellowish brown, approaching to black, with thick lips, flat noses, low foreheads, woolly hair, scattered in tufts over the head, and a pointed chin. *This race includes* the Hottentots, and savages of New Holland, New Guinea, and New Caledonia.

XXVI. Civilization.

How may mankind be classed in respect to *Civilization?* What can you say of the *Savage* class? *Barbarous? Half civilized? Civilized?*

Notes.

Mankind may be divided in respect to civilization into 4 classes, viz.: the savage, the barbarous, the half civilized, and the civilized.

1. *Those who are savage subsist chiefly* by hunting, fishing, and the spontaneous productions of the earth. *This class includes* the American Indians, negroes, and natives of New Holland.

2. *Those which are barbarous derive their subsistence* chiefly from pasturage and rude agriculture. *This class includes* the Arabs, Moors, Tartars, and Malays.

3. *The occupations of the half civilized are* agriculture, and manufactures, which are sometimes carried to a high degree of improvement ; but foreign commerce exists only in a limited degree. *The nations which belong to this class are* those of China, Japan, Southern Asia, Persia, Turkey, and Northern Africa.

4. *Nations which are civilized, excel in* agriculture, manufactures, literature, science, arts, and commerce. *To this class belong* Europeans and their descendants.

General Remarks.
I. TIDES AND CURRENTS.

1. The *tide* is the regular elevation and depression of the ocean, which occurs twice in every 24 hours. *The cause of the tide* is the

ttraction of the moon chiefly, modified in some degree, by that of he sun. *The tide begins to rise at any place*, at the rise of the noon, and continues to increase until the moon has passed the neridian, when it gradually sinks. *The tide is again high when* he moon is on the opposite side of the earth, and then falls again antil the moon rises.

2. *A spring tide* is a higher tide than usual, *caused by* the union of the attraction of both sun and moon; neap tide is a less tide than usual, and *takes place when* the attraction of the sun is in opposition to that of the moon.

3. *The tides are the greatest* under the equator, because the influence of the moon is felt there the most; and *least* at the poles, where this influence is scarcely felt at all. Local causes *produce some exceptions to this rule. In open situations*, as in the islands of the Pacific, the tides are regular, and generally do not exceed one or two feet; but *in bays, channels, straits*, &c. the height of the tide is often increased. Upon the coasts of France, which border the English channel, it rises to an enormous height; at St. Maloes to 50 feet; at Hamburg, which is 30 leagues from the mouth of the Elbe, the tide rises, ordinarily, about 7 feet; but when a strong wind prevails from the north east, it rises frequently to 20 feet. In the bay of Fundy it reaches even to 60 feet.

4. *The ocean, in many parts of the world, is subject to* currents of great force, occasioned by the tides, winds, and other causes; the Gulf Stream, p. 25, is one of the most remarkable.

II. WINDS.

1. *Wind* is air put in motion.

2. The *Trade Winds* are remarkable currents which blow from east to west, between about 28 degrees north and south of the equator, nearly round the globe. *These winds are called Trade Winds*, because they facilitate trading voyages.

3. *The Monsoons* are periodical winds which prevail chiefly in the Indian Ocean, blowing half the year northeast, and the other half southwest. *The Monsoons extend* over the whole of India and the sea coast of East Persia. The southwest monsoon lasts from April to October, bringing with it rain and tempest; the northeast monsoon produces a dry and agreeable state of the air *The cause of the monsoons* is unknown.

4. *Land and sea breezes* are winds which prevail in islands and on coasts in warm climates—the land breeze blowing towards the sea during the night, and the sea breeze blowing towards the land during the day.

5. *Hurricanes* are violent storms occurring in hot climates, in which the wind changes in a very short time to every point of the compass, with almost irresistible fury.

6. *A gentle breeze moves* about 4 miles an hour; *a high wind* 30 miles; *a tempest* 50; *a violent hurricane* 100.

7. The *Samiel*, or *Simoon*, is a wind from the deserts of Africa, Arabia, and countries neighbouring to these, so hot and pestiferous, as sometimes to produce instant death. The *Harmattan* is a similar wind which blows from the Sahara upon the western coast of Africa producing dryness and heat which are nearly insupportable.

3. The *Sirocco* is a warm and unpleasant wind from Africa which prevails in the southern countries of Europe, particularly in some parts of Spain.

III. Rain.

1. *Rain* is water falling in drops from the clouds, the *ordinary height of which* does not exceed one or two miles.

2. Mountains *receive more rain and snow* than places of less elevation; *islands, and places near the sea*, also, receive more than those which are more inland.

4. The Torrid *Zone receives the greatest quantity of rain*, beyond which the quantity generally diminishes as the latitude increases.

5. *Thunder storms are most violent* in the Torrid Zone, and less violent as you approach the highest latitudes, where they are unknown. *Rain is generally unknown* in many parts of Asia and Africa, particularly in Egypt; and in Peru and Chili, in South America. These countries, and others like them, are entirely watered by streams from the mountains, and by dews, which are much heavier than in other countries.

IV. TEMPERATURE AND PHYSICAL CLIMATE.

1. Those rays of the sun which fall perpendicularly, *heat the earth most;* hence countries near the equator have a hotter climate than those at a distance from it.

2. The *heat diminishes,* and the *cold increases,* as you rise above the level of the sea; thus, under the equator, mountains 3 miles high, are covered with perpetual snow.

3. Those countries whose general inclination is towards the sun, *are warmer* than those which are turned from it, because they receive more heat from it.

4. *Mountains north of the equator, render a climate colder when* they lie on the southern border of a country, as in Siberia; and milder when they lie on the northern border, as the Alps in Europe; to which Italy is indebted for its perpetual spring and double harvests.

5. The *heat and cold of the ocean*, are in general, not so great as *of the land;* countries, therefore, which border upon the sea, and islands, are colder in summer, and warmer in winter, than other places.

6. The *soil of a country has an influence on its climate by its* warm or cold nature, and by the nature and amount of its exhalations. Some soils acquire heat sooner, and retain it longer, than others. The kind and amount also of exhalations from various soils differs.

7. *Civilization and population exert an influence on climate,* since in countries where they exist, forests are cleared, marshes are drained, and the channels of rivers cleared, and of which tend to render a climate more salubrious.

8. *Winds exert an influence on climate;* those from polar regions tend to render a climate cooler; equatorial winds tend to render a climate warmer. Winds from the ocean reduce the extremes of summer and winter.

FINIS.

CPSIA information can be obtained
at www.ICGtesting.com
Printed in the USA
BVHW04s1227041018
529296BV00028B/712/P